A SECRET
OF THE
SPHINX

THE MacHUGH MEMOIRS
(1798 - 1801)

*Best wishes to Shafida
and her wonderful smile,
James McWilliams (Jim)
25 March, 2019.*

FriesenPress

Suite 300 - 990 Fort St
Victoria, BC, V8V 3K2
Canada

www.friesenpress.com

Copyright © 2016 by James L. McWilliams
First Edition — 2016

www.jlmcwilliams.com

All rights reserved.

No part of this publication may be reproduced in any form, or by any means, electronic or mechanical, including photocopying, recording, or any information browsing, storage, or retrieval system, without permission in writing from FriesenPress.

ISBN
978-1-4602-8386-8 (Hardcover)
978-1-4602-8387-5 (Paperback)
978-1-4602-8388-2 (eBook)

1. BIOGRAPHY & AUTOBIOGRAPHY, PERSONAL MEMOIRS

Distributed to the trade by The Ingram Book Company

THE MacHUGH MEMOIRS
(1798 - 1801)

A SECRET OF THE SPHINX

JAMES L. McWILLIAMS

"Soldiers, forty centuries look down upon you!"
- Napoleon Bonaparte

ALSO WRITTEN BY JAMES L. McWILLIAMS

THE SUICIDE BATTALION
(in collaboration with R. James Steel)

~

GAS! THE BATTLE FOR YPRES, 1915
(in collaboration with R. James Steel)

~

AMIENS: DAWN OF VICTORY
(in collaboration with R. James Steel)
Reprinted in Britain as AMIENS 1918

~

A PIPER'S WORLD
99 Tunes from Around the World
collected and arranged for the Highland Bagpipe
by Pipe Major James L. McWilliams

~

THE MacHUGH MEMOIRS
BLACK WAR-BONNET: 1795-1798
THE FUGITIVES: 1810

Acknowledgements

I wish to thank Daniel Keith Sutherland for all the art work including those credited to "R.G. MacHugh." For the amazing cover design and the maps included with the text I gratefully thank Iryna Spica of Spica Book Design. I would particularly like to express my appreciation for the intense proof-reading and insightful suggestions by a friend of my youth and college years, Victor Sutton.

I would also like to thank Craig Shemilt for his valuable support.

James L. McWilliams

For my two sons, Lachlan and Colin,
with love and extreme pride.

Editor's Foreword

Roderick Gaspard MacHugh was born in Montreal, Canada, on New Years Day, 1782. At the age of ten, while visiting relatives in France, he witnessed the murder of his mother at the hands of a Paris mob during the infamous "Prison Massacres" of September 2, 1792. From the age of thirteen to sixteen 'Rory' lived with the Blackfoot Indians on the Great Plains of North America where he earned the status of a warrior and the name "Black War-Bonnet". As was the custom, he was "adopted" by his Spirit Guide, a Great Snowy Owl, and experienced his "Vision". This Vision, which, like all Blackfoot, MacHugh took seriously, foresaw him becoming a red-coated Highland soldier. With that and revenge for his mother's murder in mind, the sixteen-year-old MacHugh, in the Spring of 1798, left the North American plains on the first ship that spring from Hudson's Bay bound for London.

This portion of "**The MacHugh Memoirs**" was discovered by Alan Turnbull, a friend in Stirling, Scotland, when a local farmstead was being converted into the popular pub, "***The Birds and Bees***" in the Causewayhead district of that historic city. The manuscript in MacHugh's own handwriting, several portrait-like drawings signed by MacHugh, and three hand-drawn maps (A, B, C, appended) were among several carefully wrapped documents preserved in a small metal box. Two scraps of what looked like handwritten codes of some kind were at the bottom. Study revealed them to be musical compositions, "THE SWEET PINE HILLS", a slow air, and a march, "OUR OLD CIAMAR A THA THU" by MacHugh himself.

As Editor of **The MacHugh Memoirs** I have limited myself to modernizing the spelling and punctuation, organizing the work into chapters, and providing glossaries of contemporary English slang and Turkish and Egyptian terms which it might benefit the reader to turn to first. Also included is an Appendix dealing with the 79th Cameron Highlanders. Where it seemed useful to clarify historical points, end-notes have been appended.

    James L. McWilliams

Table of Contents

Acknowledgements vii
Editor's Foreword ix

PART I "The Convict" 1
  1 "Welcome to London!" 3
  2 "Justice" 10
  3 "My Vision Returns" 20

PART II "The Mameluke" 37
  4 "The Great Game" 39
  5 "Midnight Introductions" 46
  6 "Settling In" 57
  7 "Recalled to Duty" 82
  8 "The Two Butchers" 98
  9 "Grenadier Jardinier" 119
  10 "Caspar's Return" 137
  11 "More Psychological Warfare" 158
  12 "Farewells" 172

PART III "The Piper" 197
  13 "A Kilted Red-Coat" 199
  14 "Reunions" 233
  15 "The Sphinx's Warning" 259
  16 "The Night of Many Reckonings" 269
  17 "A Last Tune in Egypt" 283
  Editor's Postscript 299

Glossary Of British Slang 306
Glossary of Turkish and Egyptian Terms 309
Appendix: The 79th Cameron Highlanders 312
Notes............. 314
Musical Compositions by Piper R.G. MacHugh
The Sweet Pine Hills... 328
Our Old Ciamar A Tha Thu 329

List of Illustrations and Maps

Poster: "Wanted for Murder and Escaping Justice" 35
Map: "My Mediterranean" 38
"Silence !"...54
"Roustam Raza, My Mameluke Brother" 64
"Uncle Pierre" 122
Map: "Cairo and Environs" 136
"The Sphinx"..195
"A Piper of the 79th" (a self-portrait?) 198
Map: "Alexandria and Environs" 214
"Lt. Col. Allan Cameron of Erracht,
Old Ciamar A Tha Thu" 258
"The Rosetta Stone".....................................281

PART I

"THE CONVICT"

London, England
May - September, 1798

CHAPTER 1
"Welcome to London!"
(London -- May, 1798)

"Welcome to London, dearie'! Ooo, ain't you the sweetie!" cooed the young wench as she slipped her arm into mine. Not recognizing this as professional patter, I took it as a compliment. Swaggering along the London docks togged in my best buckskins and moccasins I fancied myself quite a swell. I had yet to realize that I was as much out of fashion as if I had worn feathers and war-paint. "Like to dance the blanket hornpipe wiv me, darlin'?" she continued. "I'll make you forget them months at sea -- if you 'ave the quid."

She was sort of pretty, but in a slatternly way -- and she smelled! She couldn't hold a candle to Lark's Wing, but I reminded myself that I had resolved many times every day to forget Lark's Wing. This little ragamuffin wasn't my style -- although she was pleasant and did have good taste in men. Not wanting to be impolite, I chuckled and tried to extricate my arm. "Thank you, miss, but I have to get about my business, eh?"

Two ship-mates passed by at that moment, and they laughed uproariously. "Some bloody 'business' that! Aye, blowen, snag him while you can," called Harry Clayhorn. "Young Rory's set on become a friggin' lobster."

"You'd best do all the swiving you can, Rory," yelled his companion, "before the Frenchies splinter your mainmast and blast away your tackle! Har har har!"

Their laughter attracted more 'ladies', and mercifully my comrades were immediately distracted by petticoats and cleavage. The Cockney girl, however, grew more insistent. "Come wiv me, dearie! Katy'll make you fergit all that there ice and snow. I know you just arrived from Rupertsland and will be wantin' some lovin', and I'm just the woman for a fine swell like you."

"Woman!" I laughed scornfully, feeling very adult for my sixteen years, "You're just a slip of a girl! No thank you, miss Katy. Although you're very pretty, I really must be about my business."

"So you're joinin' the Army? What a pity! There won't be any girls there, love. Come on! I'll show you a time to remember!" By now she was pleading, and I was feeling sorry for her. I noticed a weasel-faced young character leaning against a wall, watching us as he picked at his teeth. Throwing down his dental implement he pushed through the throng of sailors and whores -- for I now recognized the "ladies" for what they were. My crew-mates had talked incessantly about "whores" for weeks, but I had expected something more glamorous than this gaggle of trollops.

"Please, mister! I'm still a virgin -- but I'll be real good!" Weasel-face was almost beside us now. "Please, sir! Come wiv me now! 'ee'll beat me if'n you don't. It won't cost you 'ardly anything!"

"Hey, bitch!" snarled Weasel-face, "we're not givin' it away! I told you to get an officer -- one what 'as silver and'll pay top quid fer a virgin." He turned to me. "What's wrong, Doodle? You a gelding? You've got the silver, I know, for you ain't 'ad no time to spend it. You've just stepped off that there 'udson's Bay ship. Don't worry none about your tackle. This dell's no fireship. Katie's a clean little piece -- and a virgin."

I must have looked disgusted, for he turned from me and bawled, "Any you gents fancy un-corkin' a new bottle? Got a great little dell 'ere, un-peppered and fit as a guinea. Yer tackle's safe 'ere, lads. But it'll cost ye!" Spying some ship's officers, he roared, "Sirs! I've got just the lamb 'ere to change yer eyes to saucers and yer bowsprits to mainmasts. Never poxed, and a virgin to boot!" He turned to push the girl forward, but Katie stood looking at me. "Dammit, blowen!" he hissed, "Come when I call, you silly little slut!" and he slapped her face.

I responded for her by slapping his face so hard that I drew tears from his shifty eyes. He staggered back, holding his cheek. "Learn how to address a lady, mister," was all I could say.

Holding a hand to his flaming face, he glared at me. Suddenly a knife appeared in the other hand. "You little bastard, you're dead! **Nobody** lays a 'and on Taddy Fromm!" and he lunged at me agile as a cat. Amid shrieks, the whores and sailors jostled back to leave us room to fight.

Taddy was quick and used to street brawls. I suppose he was expecting easy pickings, for after all, I was only sixteen and probably didn't look much older. He couldn't know I had already fought and killed three men, each deadlier than him. Nevertheless, his over-reaction had taken me completely by surprise. His knife darted out, but I jumped back. Three years of training resurfaced as my mind dismissed all thoughts except survival. I remembered Sun Bull's words, "When threatened by any animal you must attack." I crouched and circled Taddy, steadily forcing him back. Although he held the only weapon his confidence began to wane for he could see I relished the game he had started. I was about to rashly throw myself into a struggle to take his knife, when Taddy panicked and slashed at me in a round-house swing. I let him come close, but when his arm passed I seized his wrist and turned it sharply up behind him. With a scream of pain and terror, Taddy dropped the knife, and I threw him face first onto the cobbles then straddled his back. The blade glinted in the watery morning light, and for a moment I was Black War-Bonnet again. Taddy's knife seemed to leap into my hand as I yanked his head back by the hair. I held the point to his outstretched neck.

"My God, the lad'll top 'im!' exclaimed someone.

A girl screamed, "Don't kill 'im! Please!" and threw herself in front of us. It was Katie. "Please don't kill 'im, sir. I love 'im!"

I came to my senses and threw down the knife, let go of his hair, and stood up. "He's all yours, miss, but you deserve better." I made a grand bow, but a blow to the back of my head ended my venture into chivalry.

~

When I came to I was lying on the cobblestones beside Taddy Fromm. Our hands were manacled behind our backs. The young pimp was shouting obscenities at a large, bulldog-faced man holding a truncheon. Taddy continued ranting until a boot in the crotch cut short his tune.

"I'm arrestin' both of you fer disturbin' the peace in my parish -- and *you*," he added, pointing his truncheon at me, "will be tried fer attempted murder. Fromm, you slimy little Flash Man, you'll be sent up fer consortin' wiv whores and causin' public disorder." With that he hauled us to our feet, Taddy still bent over and whimpering. From a safe distance the crowd, which had grown since I last saw it, growled in discontent.

Being barely ten minutes ashore, I had no idea where we were taken, but we ended up facing a magistrate of some sort in a room labelled "Public Office", in what appeared to be a tavern. It was soon crowded with noisy spectators.

"Order, I say! Order!" shouted the magistrate, a jaunty sort whom I can still only think of as a 'chap'. "There, there," he admonished the crowd jovially, "settle down and let us get on with the business of this court. Beadle Ryfet, I see you have brought us a desperate pair to punish for their sins.[1]

"I say!" he exclaimed, looking at me for the first time, "I do believe this one must be a Yankee! It's obvious, lad, by your strange mode of dress, that you are not from here. T'is a shame you are of a criminal bent, for I would like to interrogate you regarding life in the wilds amongst the savages of our late-lamented colonies. Pray tell, are you acquainted with any of the savage redskins that inhabit your homeland?"

"Yes, sir, but I am not a Yankee, your Honour. I am a British citizen."

"Ah! Then by your own admission you are a rebel **and** a traitor, a British citizen who resides in the so-called 'United States of America' and consorts with bloodthirsty red heathens. Quickly, sir, your name! My court does not have all day, even in a case of treason. I have my writing to attend to. Quickly, your name!"

"Roderick MacHugh, sir," I mumbled. "But I'm Canadian, sir, not a Yankee." Apparently interested only in my wardrobe, he ignored my words as he busily penned something on a sheaf of papers. "I'm no traitor, sir," I trailed off lamely. It seemed best to say as little as possible before this magistrate.

"Ah, yes. You should have made that clear in the first place, Hugh," he said as he concluded his writing with a flourish. "I would not wish it said -- even untruthfully -- that a foreigner appearing before my court was unjustly dealt with. Now, Beadle Ryfet, with what are these two miscreants charged?"

"Yer 'onor, I found them a-fightin' on the docks. The youngest one 'ere, sir, 'ee 'ad 'is knife at the throat of the pimp, yer 'onor. It would 'av been as well fer the peace of the parish if'n I'd let 'im finish off the slimy little bastard, but I seen my duty, yer 'onor, an' I done it."

Taddy Fromm had regained some of his color and cockiness, for he blurted, "Ryfet, you bastard, you just want to swive my sweetheart, Katie, sweet innocent virgin that she is. You want to sell 'er on the docks, yerself. You've been after my girl all along!" There was a growl of general agreement from the spectators.

"I say!" admonished the magistrate, "we can't have public servants reviled in such terms, young man. If you persist, I shall be forced to deal with you very harshly indeed." He then turned to the spectators and asked, "Were any of you witness to the event described by Beadle Ryfet?"

Several stated that they had been there, and corroborated the beadle's story, although my shipmate, Harry Clayhorn, insisted that the knife in question was not mine, but had been drawn by my opponent.

The magistrate cut short all the so-called evidence. "First I must commend Beadle Ryfet for his diligence and courage in keeping the streets of our parish free from the criminal element. You, young man," and he stared severely at me, "are a disgrace to your new country, and living testimony of your countrymen's folly in cutting themselves off from the civilizing influence of the motherland. I would have liked to interview you regarding the customs of your heathen redskin friends to add even more verisimilitude to the monumental history which I am preparing for publication." Benignly scanning the crowd, he added, "No doubt many of you here present will wish to avail yourselves of this scholarly work when it becomes available to the general public." Then turning to me he continued, "However, we must all make sacrifices for the public good. So I will forego the fruits of such an interview in the interests of public safety. I hereby sentence you both to twenty years of penal servitude for attempted murder." This was received by the spectators with a mixture of agreement and dismay. Katie, the alleged virgin, shrieked in anguish.

"As we all know, due to the increase in criminals apprehended by the diligence of our men of the Watch, and of course our dedicated beadles, the prisons are filled to overflowing. Consequently, I have no alternative but to order you both removed to the hulks that lie off Woolwich. May God have mercy upon you. You will need it."

CHAPTER 2
"Justice"
London, May - July, 1798

The "hulks" are the true Hell on earth. I cannot believe that any nation, let alone a bulwark of civilization governed by supposedly upright, God-fearing citizens could send even the worst criminals to live in the hulks. That they send youngsters is beyond belief. The hulks are the remains of once-proud ships dismasted and anchored off-shore to serve as floating prisons. Living conditions are the most appalling I have ever seen, with hundreds of men of the most desperate character jammed into a space once designed to accommodate a few cannon and several dozen sailors. Disease and pestilence sweep through these floating hells, carrying off an incredibly high percentage of the unfortunate inmates to be unceremoniously dumped overboard into the Thames.[2]

I was taken, along with the pimp, Taddy Fromm, to be turned over to the guards on *Justice*[3], an ancient warship of moderate size. I have no idea how many of us were quartered there, but I do know that if we had all lain down at the same time to sleep, there would not have been enough floor space to accommodate us. As it was, I never once lay down there.

Every night I slept ever so lightly, seated with my back against the bulwarks.

As we passed through the gates of the Dockyard at Woolwich my reluctant companion began to tremble. "God in 'eaven, they'll kill us there! We'll be rogered to death cuz we're young and small! I'll jump overboard and drown first. The bastards'll not bugger me!"

I was green as spring grass, but even I had heard of such things on my voyage from Hudson's Bay. Aboard ship there was much ribald humor about sex between men, but here was a ruthless young villain terrified out of his wits. It made me think. "Our only chance is to team up," said I more confidently than I actually felt. "If we stick together, and take turns staying awake, we'll get by. Shake on it?" I asked, offering my hand. He looked me over, then seized my shackled hand.

While passing a ramshackle graveyard, we caught our first glimpse of *Justice* painted, appropriately enough, in black. It was unlike any ship I had ever seen, for it bore only two stunted masts which served as clothes-racks for the tattered rags hung out to air by its inmates. *Justice* was a monstrous place. Words fail to convey the depravity, the filth, or the stench. Packed tight with vermin, human and otherwise, it is what Hell must be like. I have spent more than my fair share of time in prisons and I have endured some of the bloodiest battles in history. I have been severely beaten, and I have all but frozen to death in prairie blizzards, but every one of these ghastly experiences pales beside my memories of *Justice*.

Our shackles removed, Taddy Fromm and I were hustled below decks just as our fellow convicts were returning after their day's labor ashore. We were led down three flights of stairs to the bottom deck. Newcomers, we learned, started at the bottom, and if they survived and if their behaviour warranted it, worked their way up through the decks. Taddy and I found a small cranny where we could sit together and be more or less surrounded on three sides by the bulwarks. Immediately the deck began to fill with some of the vilest men I have ever encountered. I shivered and shrunk back into my spot trying not to attract attention. Presently an ominous silence fell over our little refuge. I looked up into the hard, unsmiling face of a convict of medium size and unremarkable features.

"You're in my spot," he said quietly.

I had been prepared to resist, but something about this man warned me to back down. I pushed Taddy to the side and shuffled over. The quiet man sat down wearily beside me and stretched his legs.

"Now you're in my spot," declared a tall skinny fellow affably enough.

"I'm sorry, but I was here first," I declared.

"Like 'ell you was, sonny. I bin 'ere fifteen bloody months, and that's **my** spot."

Nonplussed by the sheer logic of his argument, I made to shuffle over again when another convict with stringy blond hair and wild eyes grabbed my arm and jerked me to my feet. As he moved to take my place I seized his arm in turn and spun him around. "Bugger off, mite!" he snarled and swung wildly at my face. I ducked and seized his wrist as it flew by, turning my back into him and throwing him over my shoulder. He crashed to the wooden floor and slid several feet on his stomach. He was up immediately, berserk rage suffusing his pitted face. I was shocked by the madness in his eyes. Without a word he threw himself at me again. I stepped back, but someone put out a foot and tripped me. As I stumbled to the floor I instinctively rolled aside. My assailant dived but missed me and I was able to snake up on top of him and grab a mass of tangled hair. There was little space to manoeuvre; he would soon regain his feet. I whacked his head on the grimy deck a couple of times. Jerking his head back a third time, I asked quietly, "Do you want your skull smashed?"

"No!" he gasped. "No more!"

I jumped to my feet. He must have been winded, for he lay there gasping for a moment. I crouched, half expecting another attack, but no one moved. In fact, only a few who were not already seated showed much interest, and of them half were grinning.

"Looks like the soft spot on the planking is yours," came the voice of the quiet fellow.

" 'ope we're not all going to 'ave to fight you for our reservations at the swill trough as well," chuckled the tall skinny man.

I sat down cautiously as my opponent rose from the floor snarling once again, although not advancing. Taddy moved over to make room for me, his weasel face pale as a fish's belly.

"What's your dodge?" asked my tall neighbor, drawing his long legs up to hug them and stretch his back. Seeing my confused expression, he tried again. "What're you in for? What's yer game? I can see you're no Rampsman, and I doubt you're a Sneaksman[4], so why are you 'ere?"

"We're murderers," declared Taddy Fromm. When those about us started to laugh he realized he had over-played his hand and tried to back up without losing too much face. "I mean we're 'ere for attempted murder -- both of us."

"Ha ha! Whose little grey granny did you try to top? One of your whores? You can't gammon anyone 'ere, kid. You're a bloody pimp -- or at best a diver or someone's shill."

"What about you, kiddy?" asked the skinny fellow looking at me. "Who'd you murder?"

"No one," I declared. "We were just fighting -- me and him -- and the Beadle claimed it was 'Attempted Murder'."

"And a dangerous cove you are, I'll wager," he said with a laugh. "Got a name?"

"Rory MacHugh," I muttered, ashamed to mention my name in such a place.

"A bloody oatmeal-savage -- though you don't talk like one. Sounds like a bloody Yank to me. What do you think, Ned? Should we call 'im 'Yank'?" he asked the quiet man beside him.

"Aye, Shorty, call 'im 'Yank' if'n you like," replied Ned Furney, the famous cracksman, with little apparent interest. From then on I was known as "Yank" in London's criminal world.

In a few minutes we were instructed to set up long dining tables in the hold to receive our evening meal, a pint of slop magnanimously referred to as 'gruel'. Supper invariably consisted of a crust of bread and gruel -- supposedly a measured pint. Curled up on our hard-earned bit of deck, the former pimp and I listened as the others sat around and talked. The order was given to turn in -- allegedly to enjoy the sleep of honest labor. However, we convicts were now on our own, as the hatches had been battened down with only one warder left on guard above decks. The authorities considered us secure enough, locked into our floating Hell-hole moored in the Thames. There was no apparent regard for our safety, and I often wondered how they explained away the possibility of fire, for we would have all been burned to death if a fire had ever broken out on *Justice*.

Most of the convicts aboard *Justice* had been there for many years, and had developed "dodges" to improve their lots. Coin-forging operations swung into action each night as soon as the ship had been battened down by our 'screws'. Various other schemes were hatched to be put into play during the day when most of us were taken ashore to do hard labor. Aboard *Justice* I witnessed scenes of the most terrible brutality during those awful nights. Many of the deaths listed as accidental or disease-related were actually the direct results of the savagery which dominated the hulks each night. All ages and classes of criminal were thrown together, so the youthful petty-offender was given a thorough schooling in crime and degradation. Rare indeed would be the prisoner who left the hulks a reformed man.

I had learned a long time ago that to show weakness was to invite brutality, so I kept my emotions to myself and displayed a stony face to hide the sorrow, humiliation, and terror that almost engulfed me. Many nights shivering in my stained sheet and dirty blanket my eyes filled with tears as I recalled my Blackfoot life and friends, and lamented my foolishness in leaving the Lords of the Plains. 'Why had I left?' I sometimes asked myself. Deep down I knew the answer, one so impossible that I hesitated to contemplate it. I had vowed to avenge my mother's murder by tracking down 'Moses Roses'. Fortunately other memories -- my Sun Dance, my two fathers, both mothers, and all their sacrifices for me -- gave me the courage I needed to survive. I convinced myself that I would endure my present life and go on to better things, for my Vision had foretold much that had not yet come to pass. Both my fathers had been prisoners; and both had risen to become leaders of their people. They would help me survive.

Fortune had smiled upon me in one way at least. I had found myself a niche. The convicts on the lowest deck were in awe of the Cracksman, Ned Furney. Although he did not look impressive, Ned was almost as tough as my Blackfoot uncle, Kills Twice. Silent and tough, Ned Furney was respected as a star at his trade, and among thieves was considered an honorable man. He had avoided capture these many years despite being a practitioner of one the most dangerous professions -- a second-storey man. He had only been taken this time because he had given up his escape to return to the aid of his confederate, Shorty Tennant, who had been pinned by the Night Watch. Now he and Shorty reigned as King and Duke of the third deck. Many tried to associate themselves with this court by paying homage to 'his highness'. My display of bravado that first afternoon battling the demented Sammy Thirt had amused Ned, and he adopted me as one of his entourage. That made my life a lot easier than it might have been.

Poor Taddy Fromm! I actually felt sorry for the little rat. He toadied constantly to Ned Furney, but to no avail. Ned despised him. Soon Taddy turned to toadying to me, when he thought that might win him royal favour. Because we had gotten into this fix together, and because we had vowed to stick together to survive, I felt responsible for Taddy and did what I could for him. If nothing else, my friendship with the master Rampsman saved us both from the sexual perversions which befell most of the younger convicts.

Each morning we rose at 5:30, washed after a fashion, rolled up our bedding and set up the tables where we ate our crusts and gulped a pale warm liquid mendaciously described as "cocoa". We were put to work cleaning the quarters before roll-call and then marched off to do four and a half hours labor ashore. Boats crewed by trusty convicts arrived alongside and we were loaded, 110 to a boat, and rowed ashore sharp at seven o'clock. There we were formed up in military style and marched off to our various tasks under the watchful eyes of numerous armed guards. At noon we were marched back, loaded into the boats and returned to *Justice* for dinner, a 'soup' concocted by boiling a few bones with some spuds. An hour later we were returned to shore to resume our labors for another four and a half hours after which we were rowed back to the hulk to enjoy the evening bowl of gruel, clean up, and hit the board floors at 8:30. At nine we were ordered into total silence, and shortly after were 'busy as thieves' forging coin of the realm and furthering the progress of crime in a multitude of projects.

Although I make light of this life now, it is only because I can stand to speak of it in no other way. The filth and the incredible stench were particularly loathsome to me after years amongst the fastidious Blackfoot. In vain I clutched the tiny Medicine Bundle that hung around my neck and prayed each night that the Great Snowy Owl would return to guide me. I pleaded for a Vision to reassure me I had taken the right course, that somehow all this horror would lead me along the path shown me that night on the West Butte of the Sweet Pine Hills.[5] But I saw no owl, heard no piercing cry in the night except for those of my fellow convicts. The starvation diet, the lack of any intellectual activity, and the constant need for vigilance against my fellows, made life seem unbearable at times, but there was no alternative except suicide, and though many chose that method, I vowed I would find another way to escape *Justice*.

At that time we convicts were employed as slave-labor unloading and stacking timber on the docks. Hundreds of honest unemployed labourers lost the means of feeding their families due to our work although sometimes gangs of workmen were hired to come in and do specialized jobs. On such occasions they mingled with us. At first I thought I might contrive to leave with them at the end of the work-day, but as we convicts wore our ugly brown uniform with its distinctive red stripes and were formed up for roll-call and marched off before the civilians were allowed to leave, I gave up that idea.

However, it had not taken me long to realize that it was easy to pass messages back and forth through these workmen, the convict boatmen, and even several warders who enjoyed their dram more than was wise. Although we were supposed to be totally without possessions aboard *Justice*, nearly all the convicts had secreted weapons, money, forging tools, and other "necessities" aboard the rotting hulk. If such contraband could be slipped in it must be equally possible to slip something out -- me. One night, convinced I would go mad if I did not escape from *Justice*, I declared to Ned, "I am going to find a way to get out of here -- ***soon***."

At first he just laughed, but as I relentlessly explored all the possibilities, he realized I was not just one of the many who only dreamed of escape. One day he took me aside. "Shorty and I 'ave friends on the outside -- colleagues suffering grievous financial loss without the industry of the short one and yours truly. Our Fence-In-Chief has discovered the difficulty of living off 'onest commerce and is anxious to resume our partnership. I've received several messages from 'im in recent weeks, and mean to make good our escape *soon*. I tell you this, Yank, in strictest confidence, for I've decided to include you in my plans."

CHAPTER 3
"My Vision Returns"
(London, September, 1798)

"Make sure that you work alongside me and Shorty today, Yank," Ned whispered to me one morning, "and do exactly as I tell you. Today is the day." Consequently, I pushed myself into file directly behind his partner for the morning roll-call and tramped off *Justice* with Taddy Fromm at my back. He had become obsessed with the idea that Sammy Thirt, my assailant of the first day, had transferred his hatred to him, my supposed mate. Probably he was not mistaken, for Sammy was going mad. The berserk gleam I had noted in his eye when he attacked me that first afternoon, had grown ever brighter. Madness was not a rarity aboard *Justice*; it was almost as common as "jail fever" which is the name they gave to the deadly sickness caused by malnutrition, filth, and terror.

"Lord save me!" I heard Taddy's desperate whisper. Behind him I glimpsed the leering, pitted face of Sammy Thirt. He wore a self-satisfied, conspiratorial grimace which chilled my heart. Taddy was terrified. "Ee's got a friggin' knife!"

"Silence!" roared a warder, and whacked me with his truncheon.

I heard Sammy's snicker of glee as I winced under the blow. Ahead of me Shorty's lanky form betrayed no indication that he had heard anything.

As we arrived on the dock I devoted my attention to staying beside Ned and Shorty. Every time they lifted a log I was there to assist one of them. Unfortunately, Taddy was equally determined to stick to me, and Sammy too seemed always to be there. Consequently, the five of us seemed linked together by an invisible string although I suppose I was just overly sensitive, for no one else seemed to notice this phenomenon. Our squad started by stacking a particularly high pile of timber which we had begun the previous day.

"All right, you buggers," bawled a screw, "four of you up on top *now*!"

Nobody ever wanted to work on top because the footing was dangerous and there were always less men to do the labour. We all straightened up to see who would be ordered to do the dirty work when I saw Ned step forward and begin the climb to the top. Shorty followed, so I shambled out and began to scramble up the stack of rough timber. Out of the corner of my eye I could see Taddy pushing through the convicts to follow me.

"'ere you!" came the warder's shout, "I said four, not five! Get your miserable arse down 'ere!" Sammy Thirt had pushed himself forward as well!

"That's alright, Warder," came the voice of the head screw, "They can use the extra man up there."

When I reached the top I took in the lone warder who lounged on the stack with us, his blunderbuss cradled in his arm. I looked around for Ned Furney. He had already gone to the far end of the stack and appeared to be busy straightening something at his feet although I could see he was looking out towards the marshes. Something was brewing so I risked sneaking a gander over the countryside, for we had a good vantage point. To the north flowed the Thames, while beyond it, shrouded in mists, lay the marshes of Essex. Closer, and of more interest, lay the Woolwich marshes. Beyond these were woods, also barely discernable in the early morning mist. Between us and the marshes rose a dozen stacks of timber in neat rows separated by five-foot-wide 'corridors'. If only I could get to the woods, I reasoned, I'd have a chance to escape for good.

"'ere, you! Quit your gawkin' and put your back into it!" came the warder's growl. "Get this lot straightened out," he added, pointing to a few timbers lying haphazardly on top of our pile.

As I bent to the task I realized we were out of sight of those on the ground. I was shouldered out of the way at that moment by Ned. "Go to the other end," he whispered, "and be ready to make a run for it."

Ignoring the screw's bellowed orders to "get a move on", I picked my way to the end of the squared log. My heart was pumping wildly. With Shorty I bent to lift our end of the huge timber. We straightened up and gingerly picked our way over the uneven surface. At the other end each of the three convicts wore a different expression as they maneuvered the timber into place; Sammy had that demented gleam in his eyes which he kept fixed on Taddy; the pimp's face reflected sheer terror; Ned Furney appeared totally absorbed in his work. We lined up the timber and carefully set it down. I can remember the tension in my heart as I straightened up and watched Ned.

The master cracksman stood making sounds of agony, gesticulating wildly at his foot. It looked like the timber had been set down on top of it. Half turning, he appealed to the warder who stepped forward for a closer look. As he did so Ned spun about and seized the man's blunderbuss with both hands. At that same instant Sammy Thirt, an insane leer on his face, slashed at Taddy with a knife. To avoid the blade Taddy scrambled sideways and tumbled onto his back.

The warder still grasped the gun with one outstretched hand even as Ned yanked it away. Sammy looked up, and seeing the warder's unprotected belly bulging his coat, slashed upward. The guard staggered back, and Sammy was upon him stabbing over and over again.

"Come on, Yank!" Shorty's hoarse whisper broke into my astonishment. "Scoot!"

He threw himself across to the next stack of timbers. Ned Furney had already done so. Instinct took over and I gathered myself for the leap. I landed on the next stack, lost my footing, scrambled for a moment, then dashed on. Concentrate on a rhythm! Run so many paces and hurl myself over the five-foot-wide corridors then repeat it. In no time I was on the last stack. A few yards away lay the marshes.

Incredibly, the alarm had not yet been raised! I scrambled down the wet slippery timbers to the ground. Already Ned and Shorty were splashing into the marsh. There was a terrible scream of fear from above. Even as I looked up, Sammy hurtled down, crashing onto the hard ground with as sickening crack. I recognized the sound of splintering bone. Sammy rolled about screaming and clutching at his lower leg. His bloody knife lay discarded beside him, but the warder's blood had drenched his jacket. Taddy came scrambling down the stack. He spared a second to put his boot to Sammy's head before sprinting off towards the marsh.

By now shouts could be heard, and a fusillade of shots echoed off the piled timber. Those warders armed with muskets had tried long shots at Ned and Shorty. Blunderbusses were useless at that range and would be needed to control the remaining convicts. Taddy and I sloshed into the marsh. At first it wasn't too deep, and we seemed to be making good progress. I heard the crash of a musket at no great distance behind me and almost felt the whine as the ball passed between us to splash a few feet in front. The slow pace momentarily panicked me and I began to flail about then got hold of myself and pressed on. Now the water was up to our crotches, forcing us to move so slowly that it made no sense to try to zigzag. Ned and Shorty were at least a hundred yards ahead. Another crash reverberated through the mist and whined by quite close. I could hear the convicts cheering and the screws shouting for order. At least our comrades would be getting the day off work, I thought irrationally, for the guards would already be reforming the squads to return to *Justice*.

A short fusillade rang out causing me to wince. To the rear Taddy gasped and sobbed with fear. Then all was silent except for the plopping of shots in the water **behind** us. We were beyond their range! Although the fog was thicker here, I could make out the woods looming before us. Suddenly I was on firm ground, and small trees were thick before me. A tall figure materialized to my left, and I heard Shorty's voice call softly, "Yank! Over 'ere!"

I made my way to where he stood. His face was jubilant. I understood the emotion although it seemed a bit premature. What chance did four convicts in prison uniform have?

"Let's get out of 'ere," declared Shorty. "Ned's already jawing wiv our mates, I'll wager." We plunged on into the woods, and I could hear Taddy crashing along behind us. "What's that?" asked the tall one, startled. "We're being followed!"

"It's just Taddy," I reassured him.

"Blast! Get rid of the little sneak. 'ee can look after 'imself. We didn't plan on a fourth."

"I can't do that, Shorty," I remonstrated. "We're mates."

"Mates with that little bastard?" he exclaimed incredulously. "'ee ain't worth a pint o' poxy pig piss! Ned won't take 'im. 'ee'll brain the bloody cove first!"

Taddy came crashing up behind us. "Ah, there you are! Where's friend Ned?"

"Bugger off, slime-face!" snarled Shorty, then turned and dove deeper into the woods.

I followed, and behind me came Taddy calling out, "Slow down, Yank! We're safe now."

Suddenly out of the mists the image of a coach and four horses took shape. We're trapped! I made to dash back into the trees when Ned's voice called out, "Did you bring the Yank, Shorty?"

"Aye," gasped the tall one, "but that little gutter-rat followed us."

"Damn! Well, get changed. Our friends even brought a pot of water to wash our dirty dials." I arrived beside the coach and found Ned stripped to the waist busily splashing water over his grimy face. We did the same until I heard Ned snarl, "Not you,

Fromm, you bloody pimp! Get the 'ell out of 'ere, or I'll rip your lips off!"

"Ned, he's my mate," I pleaded. "I can't leave him here to be caught by the screws."

"Then 'ee can bugger off on 'is own. I ain't stoppin' 'im from 'scapin'. But 'ee ain't comin' with us, Yank. We made provision for three of us to get away Scot-free, but there's no room in the plan for this poxy little pimp."

"Then I'll have to go on my own too," said I foolishly, "for he's my mate." There! I had said it again. I hated the thought yet I couldn't leave the weasel-faced little slug to be caught. If I did, it would haunt me for the rest of my life.

The upshot, surprisingly enough, was that Ned backed down. A few moments later when the carriage rolled out of the woods it carried besides its driver three rather seedy "gentlemen" trying to comb their filthy locks into a semblance of respectability. While hidden in the box cowered one filthy convict. Our story, if required, would be of three gentlemen on an urgent trip to London who had all unknowing picked up a stow-away convict when they had stopped to relieve themselves after getting lost on the road beside the woods.

∼

All went without a hitch, and within a few hours we were safely ensconced in a back street in the western part of London. It was there that I heard how our escape had been arranged. Ned's "Fence-In-Chief" had contacted him through the laborers who worked on the docks beside us. A coach would be waiting with suitable clothing if Ned would care to chose the date and place. Almost a week before the escape, arrangements had been made for the addition of one convict to the group -- me. The reason for this was made clear soon after we were bathed and fed.

"Well, Yank, me lad," began Ned, "I noticed when you first came aboard the hulk that you were a youngster with metal and a way of 'andling yourself in a crisis. You move like a cat and are athletic. Being a bit of a lightweight makes you perfect for the profession. I'd like you to join Shorty and me. I don't know 'ow much you know about second-storying, so I'll fill you in. We go in through the second-story windows -- them that ain't barred. It requires a lofty bloke to serve as ladder, and a light-weight chap like yourself or yours truly to shinny up into the windows. Usually we works in pairs, but it seems to me that 'aving two sharp lads like ourselves rummaging through the premises will double the chances for plunder and cut the time inside by 'alf. What do you think, Yank?"

I was astounded. I had never given any thought to his reasons for including me in the escape. "I am highly honored that the most famous cracksman in London would want to take me in as a junior partner, but this just isn't my line, Ned."

"I could insist, Yank," he said seriously. After a pause he added, "I reckon I knew all along that you're not one of us, but you just seemed too good a choice not to give it a try. So I suppose this is 'Good-bye'. Shorty and me 'ave to get back to the trade. We got debts to pay off. Seein' as you're not takin' up the profession, I guess we'll pay your share too. But what the 'ell, you were never asked if you wanted in, so consider the breakout a free one on me and the short one."

There it ended. I could hardly expect the two partners in crime to shelter me now that I had repudiated their way of life. Besides, they would be going out to work the night shift and did not want anyone around who might be in the way. They both despised Taddy Fromm, and considered him a likely snitch, so the next morning the two of us were on our way.

"We can go and rough it wiv me old ma, for a spell," suggested Taddy. "She's got a bit of a 'ole in the wall over in Seven Dials. Then I'll look up Katie, and we'll get back into mort-managing. I miss the little tart, you know. She'll be workin' for someone else right now, I suppose, but I'll straighten all that out, and she can start earning *us* a little coin of the realm. You can share with me till you get a doxie or two for yerself. I owe you, Yank."

I couldn't believe what I was hearing. Did this weasel-faced little turd believe I wanted to take up pimping? Is that how I appeared to other people? I had to get my life back on the road I had chosen. Here I was an escaped convict, consorting with a pimp and being offered a share of the profits of poor Katie's degradation! Nevertheless, I walked along with Taddy Fromm, searching for a way out. I had no idea where I was. London was a lot bigger than any town I had ever seen, and aside from my weeks on the hulks, I had put in less than six hours in the city, and all but ten minutes of that had been spent in custody.

We had not gone very far before I spied a wall covered with posters. One of the newest purported to portray the likenesses of Ned Furney and Bernard 'Shorty' Tennant, "escaped felons, murderers, and well-known rogues and house-breakers". There was also a short description of the other two escapees: "One Taddy Fromm, pimp, slight build, pinched face, shifty eyes" and "One Rodney Hughes, American, slight build, dark complexion, with long black hair". We were all wanted for "Murder and escaping justice". Evidently the deranged Sammy had succeeded in killing the warder! At least they had spelled my names wrong -- obviously the magistrate/historian's contribution -- for which I was truly thankful.

I worked out a plan of action. I would need Taddy's assistance to find my way because my accent would make me noticed every time I stopped to ask directions of a stranger. I had done what I could to change my appearance. My straight black locks had been curled by one of Ned's lady-friends, and I now sported a big floppy hat and dressed like a reasonably successful labourer. At the time of my sentencing I had been given a few moments to settle my affairs. My shipmate, Harry Clayhorn, had agreed to take my few possessions and keep them at his parents' lodgings in the village of Gravesend. I had every reason to trust Harry, for he had been an honest friend, so I thought it best to leave my few treasures with him till my prospects had become a little less chancy. These included my most valuable possession, a little piece of paper, a bank draft drawn on a Scottish bank to provide me with the cost of purchasing a commission in the Army. This was a result of my father's reputation and a fortunate friendship with the Inland Chief of the Hudson's Bay Company. Someday I would take advantage of this gift, for my Vision had told me that my future lay with the British Army.

Evidence of that Army was everywhere. During our long walk across London we encountered several units of red-coated soldiers marching to the beat of fifes and drums or military bands. I loved the catchy music and stopped every time to watch. Taddy told me that these were the "Volunteers", various regiments formed specially to repel the expected French invasion. There seemed to be lots of enthusiasm for these local soldiers, but much less for the regular soldiers. The British Army had suffered a quarter-century of humiliating defeats, and its incompetence had become legendary. Conversation in the streets indicated that not much could be expected of the regular army and its block-headed generals. The local amateur lads, however, would knock the Frenchies for the count if ever they dared to invade.

I couldn't resist reading a broadsheet, and discovered that the news of the hour was Nelson's naval victory over the French at Aboukir Bay on the coast of Egypt. Apparently, a French Army under a General Bonaparte had landed there at about the same time I had boarded *Justice*. Bonaparte had thoroughly thrashed a huge Turkish army in what the General grandly referred to as the "Battle of the Pyramids". The papers were all in a sweat about the French now being able to cut off the British forces in India. My geography was a bit sketchy at the time, but I failed to see how this could be. Anyway, Nelson had pretty much destroyed the French fleet in the Mediterranean, thus easing the threat. As I greedily read the papers, Taddy tugged at my arm impatiently. "Come on, mate. Let's find me old ma."

Eventually we reached a wretched section of the city where everyone lived in the deepest poverty. Beggars and obvious criminals slunk along the narrow, filthy streets. Life here could only be a step or two above the hulks. The only notable difference was that no one was actually compelled to live in the Seven Dials. Taddy fitted perfectly into the background here, and I could readily believe he had grown up in such a place. My disgust for him became tempered with compassion. When we stopped before a filthy alley he looked back and forth to insure no one had seen us, then we slipped in and followed a stream of sewage to a stairwell.

Ascending a couple of flights, we entered a filthy room the size of a large closet. It contained nothing that could pass for furniture. A few bundles of rags seemed to be the sole furnishings. One bundle turned out to be an old hag, her face filthy and creased, her two visible teeth stained and chipped. She croaked at us, "Be off out of 'ere! This place is mine. I'll 'ave me six strappin' big sons give you the 'eave, if'n you don't get out!"

"Ma, it's me -- Taddy!" said my 'mate' soothingly.

"Taddy? Taddy?" she asked suspiciously. "You're not Taddy. 'ee's in jail -- the 'ulks."

"No, ma, I'm 'ere! I 'scaped! No jail can 'old Taddy Fromm for long!"

She looked him over more closely before croaking almost proudly, "Well, Taddy luv, you always was a canny shaver. What do want wiv me?"

"I need a place to lay low for a few days till I find Katie and get a dodge goin'."

"You're not stayin' 'ere There's not room enough -- and besides there's my gentleman wot's coming back in a few moments. 'ee'll carry on a right lot, if'n 'ee finds you 'ere. You clear out, and don't come back unless you've got somethin' for your ma. You only turn up when yer needin' somethin'. Bugger off!"

Taddy started to protest, but Mother Fromm began howling threats which would soon draw a crowd, so I pulled him out with me. I must admit to being too appalled to consider -- even for a second -- lying down amongst those filthy rags.

Outside I decided the time had come for me to part company with my "mate". When I told Taddy I was heading for Gravesend after all he seemed relieved and offered to walk part of the way with me. He would direct me to the road once we reached the Tower. If I had known London at all I would never have accompanied him, for his destination was too near the London Docks for comfort. The truth was I also felt a little guilt at having been the one to back out on our partnership, but Taddy still was not my idea of a "mate". He was no replacement for Hunts With The Owls -- or even Talks Funny[6].

We eventually approached the famous Tower of London, and I was impressed. The fact that it had stood guard there for all those centuries intrigued me. When we passed to the east of it I still had no idea where I really was for I had been gawking at the sights -- all the vendors and pretty girls. I heard Taddy utter some exclamation, but I had eyes only for a pretty little apple-seller. When I turned away from the girl, I almost ran into Taddy who was talking animatedly with a lady. Amazed, I recognized Katie, his "virgin harlot" of our first encounter.

It wasn't easy to recognize Katie because she looked almost a lady. Her shabby street clothes had been replaced by a more elegant outfit showing she had fallen upon better times, and her once-timid features now glowed with a new confidence.

"Well, I'll be blowed, Katie! You look like the bloody Queen of Sheba -- all 'lah-dee-dah'! Ain't you the right princess!" exclaimed Taddy.

"Aye, Taddy. Things 'as changed fer the better," she replied with a confidence I could scarcely believe. She didn't flaunt herself, or preen like a whore; nor did she back down from him.

"Well, things'll get back to normal now I'm 'ere," he declared grandly. "Wiv that outfit we can 'andle a better class of clientele. Now Katie, say you loves me as much as ever, and all will be forgiven. I know you didn't get that there outfit without some bloke shellin' out big. But all's well as ends well." Seeing me, he made an introduction. "Remember Yank? It was me and 'im fightin' over you what got me put away, but we're mates now so you won't 'ave to worry about 'im no more."

"Taddy, like I said, things 'as changed. I'm not your girl now. My new man looks after me and gives me nice things, and 'ee don't make me work the streets -- and I go to church and all."

"Like bloody 'ell, Katie! You was a whore, and you'll always be a whore. Your kind don't change, so don't come all uppity wiv me. Folks don't mess with Taddy Fromm. I've killed a man, and don't you fergit it, you blowen!" He pulled his hand back to slap her, but I seized it. He glared at me, then the anger faded.

"Shut up, you silly fool!" I hissed. "Do you want to get us arrested again? I'm leaving now. Best of luck! You'll need it if you don't learn to keep your mouth shut."

"Listen to your mate, Taddy," implored Katie. "Augustus is going to meet me here. If 'ee catches you, 'ee'll kill you -- or worse!"

"I'm not backin' down for no pimp named 'Augustus'," snorted Taddy.

"I'm not a whore, and ee's not a pimp! 'ee's the beadle!"

Just like a cue for a stage-play! The next thing I heard was a mighty bellow, "Fromm, you bastard!" Somehow it was familiar.

I looked back over my shoulder, and there was Beadle Ryfet, his face crimson in fury. His eyes were fixed upon Taddy and he didn't see me at first. I might have simply melted into the crowd that was gathering, but Taddy did the only thing he could for his mate, he stuck with me. "Come on, Yank," he shrieked and grabbed my arm.

Beadle Ryfet was rumbling towards us at an alarming speed, "They're escaped murderers! Drinks all around for those as catches 'em!"

We took off, with a crowd behind us. I could hear Katie's high pitched whine, "No Augustus, no! They didn't touch me. Let them go!"

I have no idea where we ran, because I followed Taddy who knew the area like the back of his hand. The crowd pursuing us dwindled, but Beadle Ryfet never let up. For a big man he had endurance -- fueled no doubt by rage at seeing his woman accosted by her former pimp. Ryfet knew the area every bit as well as Taddy, and we could not shake him. In fact he took a short cut and actually closed the distance between us, for Taddy was not much of a runner.

Suddenly I heard a sound which has thrilled me all my life. Quite near someone struck up the bagpipe! It was a simple old tune, "**We Will Take The Good Old Way**", but it sent a surge of excitement through me even in the midst of this fateful race. Around a corner we ducked -- and stumbled into a cheering crowd which lined both sides of the street. Beyond them, mounted on a horse, rode an officer wearing a tall black bonnet like the one in my Vision, while behind him came a piper, for I could see his drones and green ribbons above the mass. We bulled our way into the crowd, intending to rush across to the other side and lose Ryfet in the unexpected numbers, but there before us marched a rabble of men in civilian clothes, all nondescript but proud.

"Good luck on you!" called voices from the crowd. Three cheers rang out with applause and "Huzzahs". The marching men called back, "We'll lick the Frenchies for you!"

~ A SECRET OF THE SPHINX ~

> # WANTED FOR MURDER AND ESCAPING JUSTICE
>
> **One TADDY FROMM**
> Pimp, slight build, pinched face, shifty eyes
>
> **One RODNEY HUGHES**
> American, slight build, dark complexion, long black hair
> Rewards are offered for information or assistance in the capture of each of these felons.

It was then that I saw the Sergeant marching at the rear. He wore the typical red coat, but he was kilted! The tartan he wore so grandly was a dark green with a maze of red stripes and one yellow line. On his head he wore the black war-bonnet of my Vision! He was accompanied by several dashing Corporals. These were the soldiers my Vision had shown me!

Leaping forward into the street, I threw away my tell-tale hat and pushed into the crowd of recruits. I had buried myself in the midst of the group when I became aware of Taddy Fromm beside me, slouched down, keeping as low as possible. "Bloody marvelous, Yank! A bloody marvelous dodge!"

The men about us looked hung-over at best, or very intoxicated at worst. But we shambled along steadily enough to the brave ringing of the pipes and the cheering of the spectators, everyone a patriot for the moment. I caught a glimpse of Beadle Ryfet searching beyond us looking for the escapees in the throng across the street. Just then beside me the Sergeant materialized, as smart a soldier as I've ever seen. His gear literally shone, not a stitch was out of line. He reached out a welcoming hand first to Taddy, who responded by staring into his hand with astonishment. My turn came next, and when the Sergeant pulled back his hand I was left holding a shilling. "Welcome to the 79th Cameron Volunteers, lads!"

My Vision was coming true!

PART II

"THE MAMELUKE"

The Mediterranean, Egypt & Syria
October, 1798 -- August, 1799

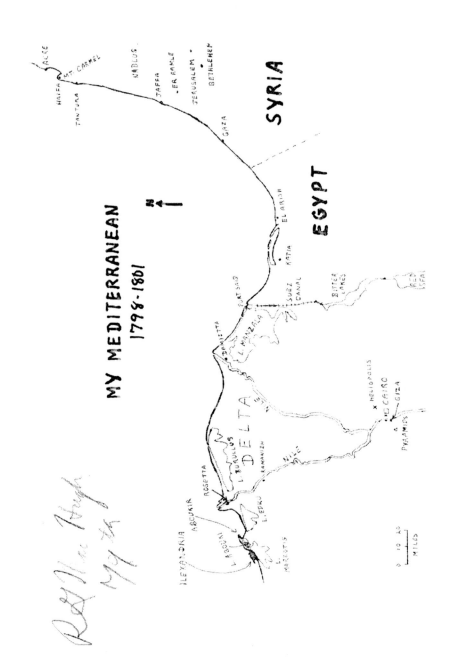

CHAPTER 4
"The Great Game"
(The Mediterranean, October - November, 1798)

A lot happened in the four months after Taddy Fromm and I enlisted in the 79th Cameron Highlanders. Being the tiniest cog in the huge machine called the British Army, I cannot even now give a clear account of what took place immediately after accepting the King's shilling but a few things stand out. It all started much as I had expected. I marched south out of London with our band of recruits bound, we were told, for the Channel Islands, where we would join the 79th in Guernsey. It soon became known that I was a piper, and I took advantage of the regular binges of our own piper to take up his pipes and play us many a mile through the countryside of Southern England. I don't know which great cogs meshed with which lesser cogs, but before our band of recruits sailed from Portsmouth to Guernsey, the gears churned out orders with momentous results for the tiniest cog of all -- me, Private Roderick MacHugh.

The Captain in charge of the Recruiting Party informed me that I had been temporarily assigned as piper to the hero of the hour, Sir Sidney Smith. This appeared to be a giant step upwards for me. Sir Sidney was a naval officer who had recently escaped from the famous French prison, The Temple, and had become the darling of the press. As a reward, Smith had just been given command of **H.M.S. Tigre**. It later transpired that Sir Sidney had also been given secret orders appointing him "joint Minister Plenipotentiary" with his brother, Spencer Smith, at the Court of the Ottoman Emperor in Constantinople.[7] Sir Sidney thought it would be appropriate for his brother to put on the dog a bit by having his own piper to strut before him and impress the Turks. It probably would have been a great dodge for me, for I love piping and I have since developed a love of that part of the world. However, neither Sir Sidney nor I had taken into account the incompetence of the British Army. I was rowed out to **H.M.S. Tigre** as it left England for duty in the Mediterranean. I had been equipped with neither uniform nor bagpipe. When I was called before the Captain, Sir Sidney looked me over, his eyes settling at last on my scruffy old jacket, a survival of my escape from *Justice*. "Ah, a member of the Royal Guttersnipe Fusiliers I presume."

"No, sir. The 79th Cameron Highlanders, sir." I knew what he was getting at, but felt it best to keep my mouth shut. I had learned at least one thing in my first weeks as a recruit.

"Doesn't the 79th wear kilts and red coats?"

I could see where this was going so I decided to get him on my side. "Yes, sir, but our company of recruits were never issued anything to wear. We weren't even paid, sir."

Sir Sidney, normally a very genteel man, fired off a few choice observations regarding the Army and its efficiency then dismissed me below deck to the tender care of the bosun. There I was regarded as "live lumber", the sailors' term for a "lobster", or red-coated soldier, aboard a ship. But even that lowly status was short-lived, and I was unofficially "pressed" into the Royal

Navy until I could be sent back to my regiment. That had been in October of 1798.

Since then I reaffirmed that my natural skills do not include seafaring. My time as a convict on the hulks had been a blessing of sorts for it had prepared me for the life of a British sailor. Had I come directly from the untrammelled freedom of a Blackfoot warrior to the lower decks of **H.M.S. Tigre**, I would have been an intolerable thorn in the side of my mates and my officers, and my lot would have been one long round of lashings at the grate. Fortunately **Justice** had taught me humility, caution, and acceptance of situations I could not alter.

Also aboard **H.M.S. Tigre** was a strange coterie of Frenchmen. At that time I cordially hated all Frenchmen, because of memories of my mother's murder. However the "Frog-Eaters" aboard **Tigre** were the ones who had carried off Sir Sidney Smith's escape from the French prison, so they were obviously men of character and enterprise, but that was not the way I saw them at first. Fortunately, despite my foolish prejudice I became involved in conversation with one of them, Count Louis-Edmond Phélippeaux. Even I immediately realized this was a man of extraordinary character, and considered him to be the ultimate aberration -- an honourable Frenchman.

In a way, Phélippeaux was a typical French aristocrat of the Old Regime -- flamboyant yet courageous modestly underplayed. As a youngster he had shared the same desk with Napoleon Bonaparte at the *'Ecole Militaire*, and cordially despised him.[8] Phélippeaux had fought against the French Republic as a Lieutenant-General in Condé's army, and had then served as a British agent for several years. It was he who had engineered Sir Sidney's recent escape -- in true Gallic style.

Phélippeaux had insinuated himself into the scene by making love to the jailer's daughter, and had exited by carrying to the prison a forged order to take Smith and Midshipman Wright to another prison. The ruse had worked, and he spirited them both out of France after another series of adventures. Smith later rewarded Phélippeaux by obtaining him a commission in the Turkish Army. The flamboyant British Captain and the aristocratic French Count had jointly sworn to harass their common foe, General Napoleon Bonaparte.

Count Phélippeaux advised Smith of my linguistic 'talents' so when we anchored at Gibraltar I was transferred to another ship where I was turned over to a civilian, Mister James R. Flemming,[9] to act as his "dog's-body". "You seem an alert lad, with a natural disposition for this kind of work, MacHugh," Flemming had remarked by way of introduction. What "this kind of work" was I did not discover for some time. Possibly I will never know the entire picture.

It began with day-long "language lessons" utilizing various seamen and locals from lands we passed by on our way to the Levant. It was all very informal and haphazard, but interesting. In a short time I was tasked with translating for Flemming when strange characters came aboard late at night. There were many such visits, and it was obvious that espionage was Flemming's occupation and passion. Many were the times he would himself slip ashore veiled in some disguise. After several hours or days he would return quite pleased with himself. Finally he popped the question that deep inside I was hoping he would ask. Nevertheless, it came as a surprise.

"You seem an alert lad, with a natural disposition for this kind of work," Flemming had remarked. How would you like to try your hand at 'The Great Game'?" I suppose I looked as confused as I felt, for after a long silent pause he explained. "I am referring of course to simply going ashore, mixing with the locals, picking up information, acting as a courier and reporting back to me. Nothing overly dangerous. Should be quite rewarding for an inquisitive and adventurous lad like yourself."

I don't know how much I believed his job-description, but it did sound intriguing. Being ashore away from the constant motion of a ship was reward enough -- and add to that the fascination and romance of mixing with the locals -- how could I do anything but accept -- immediately?

"Well then, MacHugh, let's get started. There is a lot to learn if you are to be a success at this 'Great Game' as some call it. A very serious Game it is, never doubt that." He followed this with some basic instructions on becoming invisible. 'You will often be seen but must not be recognized. No one should be able to describe you except in the most general terms. Your face is very basic, without special features. That is good. Make sure you do not develop any mannerisms. A man's walk is his most obvious visible give-away. So be conscious of that, and look for it in others. If you must be in the limelight for even a moment or two, exhibit some quirk or item of dress that you can immediately abandon when returning to obscurity. That mannerism or item of dress will be what most observers will remember. Without it you will again become invisible to all but the most observant."

I learned a lot during those weeks with Mister Flemming, but one lesson still stands out. "MacHugh, can you tell when someone is lying to you?" he asked me one evening. Of course I admitted that I could only guess.

"If you are observant and watch an individual's mannerisms you can often, or maybe even **always** detect when they are lying. We all have our own what I call 'give-aways'. One of the most common 'give-aways' I have observed is that many of us put our hands over or at least close to our mouths when we tell an untruth. It's almost like some inner moral control tries to prevent us from uttering the lie. Watch for that quirky behaviour, lad, and when you spot it, look for the lie. Other individuals hesitate or even take a deep breath before speaking a lie, almost as if they are preparing for an ordeal. Another 'give-away' I have observed is that a liar usually cannot look you in the eye while he lies to you. He, or she, will usually fix the eyes on something else while telling an untruth. Watch for this avoidance of eye-contact, lad, and you will catch quite a few lies being knowingly passed about."

For the next few days I did so and studied the sailors while they told stories of their romantic adventures or tales of derring-do. Hands were active about the lower face, and glances into space were quite common during such sessions. I really was learning to detect lies and liars.

But my newly acquired mentor was not finished with my training. "Of course, lad, you understand that we are all conditioned to behave in a like manner. When you have to tell a lie -- and you **will** be forced to do so in some of the situations you'll encounter -- you must not perform any 'give-aways' of your own. *That* is the greatest danger of all in this business," he concluded -- then thought for a second or two before adding, "well actually the second-greatest."

"What is the greatest danger then, Sir?" I asked.

"The fair sex, lad, the so-called 'fair sex'. Avoid them, MacHugh, at least until you are much older and have gained more experience in life."

There followed during the next few sessions a contest wherein we both related true stories of our pasts or of some event we knew about, but inserted a lie or two into the narration and tried to get away with it. But at the end of each story Flemming could list every lie I had told him, while I was able to detect only one or two of his. However, during our final session neither of us was able to detect many of the other's untruths. I recognized this as a sign of success. For what purpose I could not understand at the time, but I had become a good liar.

Looking back upon that week, I could now see the learning path I had been led along without even being aware of it. Four basic guidelines -- No! *rules* -- are now obvious. ONE -- Get used to being totally independent. TWO -- Pretending to be someone else means becoming that character in every conceivable way, including things you may despise. THREE -- stick to things you know (which seems to be in conflict with TWO but is equally valid). Failure is the usual result of pretending familiarity with people, concepts, or skills that are not part of your real self. FOUR -- Recognize **and use** the 'give-aways' of lying. "Keep these four simple rules in mind and you should encounter few problems you cannot handle," concluded my new mentor. I noted that he made no further mention of our "greatest danger".

CHAPTER 5
"Midnight Introductions"
(Egypt, December, 1798 -- January, 1799)

"She is so beautiful -- like a tiny doll!" exclaimed Lieutenant Fourès. "My Bellilotte is a rare jewel. Can you believe she accompanied me with General Bonaparte's Army of the Orient, dressed as a soldier of my regiment? She loves me so! I cannot believe these lies that she has betrayed me with another!" Crouched in the boat, the poor Gascon displayed none of his homeland's famous swagger. Instead bewilderment had overcome him for the last few days as his mind flitted around a circle of passions -- love, confidence, anger, dismay, and back to love.

Oblivious to their passenger's emotional outbursts, **H.M.S. Lion's** boat-crew put their backs into it and pulled us towards the shores of Egypt, an indistinct pale line in the distance. The moon was a round lantern focused upon our solitary little boat. I shuddered as I imagined hundreds of hostile eyes following us in.

"I want you to tag along with Lieutenant Fourès as his servant, a French cabin-boy, captured and taken off a prize, '**Le Corbeau**', earlier this year. You will go by the name 'Gaspard Genereaux'." His choice of name surprised and impressed me. He had obviously done some digging to discover this family name of mine.[10] "Your French is excellent, so he should not become suspicious. To explain your accent just admit the truth that you came originally from Quebec. Keep an eye on Fourès and get back to let us know what has transpired. Hopefully, he'll do the right thing and avenge himself on Bonaparte. You need do nothing but remain in the background as an observer. We have several operatives there so you won't have to make any decisions yourself, and one of them will arrange your return to the ship when there is something to report."

It seemed to make sense at the time. Months under the Mediterranean sun had darkened me so that, except for my blue eyes, I could pass equally well as a Blackfoot or as an Arab. The fact that my grasp of Arabic, the language of the masses, was still almost infinitesimal had not prevented Flemming from employing me. "That doesn't matter. You are supposed to be French, not Egyptian. Besides, with your natural language skills you'll pick it up in no time," was his off-hand conclusion.

Mister Flemming had delivered me his final instruction as I boarded the row-boat for the Egyptian shore. "My operatives will contact you by using the sentence, 'The nightingale's song is so sweet.' Your response will be, 'Yes, but only in the summer.' Now farewell and good luck, Gaspard Genereaux, my lad," and he had actually gripped my shoulder for a moment.

～

"My sweet Bellilotte would never betray me for another," Fourès asserted, having arrived once more at 'confidence' in his circle of emotions. "It will be so wonderful to embrace my darling again." The poor Lieutenant fell silent for a moment. He believed I was French, and I have always been seen to have a sympathetic ear, so he had babbled on to me ever since our introduction. Fourès wore his uniform proudly, that of a Lieutenant in the **22eme Regiment de Chasseurs à Cheval**[1], while I was clad in the non-descript pyjamas of an Egyptian peasant. On my feet were a pair of ridiculous red slippers. Their curled toes made me cringe with embarrassment, but I suppose there was a limited supply of Levantine footwear aboard a British man-of-war.

"Damn!" Fourès suddenly exclaimed. His dreamy demeanour had been replaced by rage. "I refuse to believe these lies I have been told! Bellilotte would not betray me for another -- not even General Bonaparte!"

Fourès had been taken prisoner aboard the French sloop, **Chasseur**, shortly after it had set sail. Mister Flemming had previously received intelligence of the Lieutenant -- and even more concerning his wife, the reputedly lovely 'Bellilotte' -- and had determined to capture him. I had been aboard a tiny fishing boat with Flemming when the report arrived from Cairo, and we had rushed to catch up with **H.M.S. Lion**. The influence wielded by Flemming must have been enormous for he was able to have **Lion** put in pursuit of the French sloop. **Chasseur** was too poorly armed to resist capture, so the unfortunate crew had been taken on December 19, and sent to rot in a Turkish prison, while the courier, Lieutenant Fourès, was wined and dined aboard **Lion**. Flemming, posing as a new-found friend, had revealed the trivial content of the four "secret dispatches" carried by Fourès. As Flemming had known all along, they contained only mundane reports requiring no special courier. Fourès had been dismayed, because General Bonaparte himself had entrusted him with the dispatches, having assured the Lieutenant that their delivery was of the utmost importance to France. What Flemming knew and

Fourès was slow to acknowledge was that his lovely 'Bellilotte' had attracted the eye of General Bonaparte who had concocted the dispatches to send the young husband to Paris -- and well out of the way. Just how Flemming had handled the poor cuckold I don't know, but the result was that Fourès -- the recipient of a very generous parole from the Royal Navy -- was about to land back in Egypt anxious to return to Cairo and discover the truth, and -- if it proved necessary (hopefully) -- avenge his honour by killing his Commander-in-Chief.

 The boat scraped onto the beach and out we jumped. It was a wonderful relief to stand once again upon firm ground even if it was enemy-occupied. Hopefully I searched the darkness for a sign from my Spirit Guide, the Great White Owl, but there was no sign of any kind. Within moments we were whisked off the beach by several figures in flowing robes. Being unused to the language and the geography, I know nothing of the connections made that night, but we eventually ended up at French headquarters in Alexandria. There the Lieutenant was welcomed with dismay by Bonaparte's closest friend, General Marmont, who tried his best to keep the returned husband in Alexandria. Fourès was given a comfortable apartment, and as his only "servant", I had to look after it. Our short stay gave me the chance to become acquainted with my role of servant, and I dealt for the first time with Egyptian tradesmen and shop-keepers. That was an experience! At first I was overwhelmed by their extravagant claims and emotional appeals, and I was afraid I was squandering the young officer's money by my failure to grasp the finer points of bargaining. Fortunately, I later discovered that Bonaparte's "Army of the Orient" had been footing the bill for the Lieutenant's stay.

My first venture onto the streets of Alexandria certainly dazzled me. The almost naked state of the poorer class of Alexandrians made me gawk in disbelief. Even the Plains Indians wore more! Certainly the lack of clothing on the scores of attractive young women proved very distracting to a naïve Canadian boy. Many tried to imitate the upper class ladies as best they could by stretching a bit of fabric over the lower portion of their faces even if what I considered to be more intimate parts remained open to view. Of course, I was almost equally struck by the extreme measures taken by the upper class women to make themselves almost invisible by draping their bodies and faces in great swaths of fabric.

Despite Marmont's attempts to keep him occupied far from Cairo, Lieutenant Fourès remained determined to rejoin his beloved Bellilotte and was at first bewildered by the array of excuses his superior gave for keeping him in Alexandria. Befuddlement soon turned to anger. Consequently, within a few days the Gascon had arranged for a pair of seats in the new coach service from Alexandria to Cairo, and we slipped away -- military discipline be damned. For the first time I understood the expression "taking French leave". I must admit that I found it exciting to think that once in Cairo I might actually see or possibly even meet the most famous hero of France. If Fortune smiled upon me I would have my first-ever glimpse of *"le premier l'homme de l'Europe"* -- General Napoleon Bonaparte!

Upon our arrival in Cairo late one evening, we went straight to Fourès' apartment. The love-struck husband rushed inside while I waited outside. Moments later he emerged in a blazing rage. He had been told that his loving wife had moved to Sheik El-Bekri's palace.

It was midnight when Lieutenant Fourès barged into the Sheik's huge mansion. Fortunately he kept his wits about him long enough to claim to be returning from his mission for General Bonaparte. I trailed in his wake as he stormed down various corridors, all deserted by this time. Eventually at the far end of a hallway we spied a solitary figure in baggy red pants and a turban of some sort -- one of the Mamelukes about whom I had heard so much. Fourès strode down the corridor towards the door. The young guard stepped before him to block the way. In true Mameluke fashion, he was armed to the teeth with archaic weapons, but they didn't faze the angry husband who brushed past him shouting to me, "Genereaux, take care of this fellow!"

The bewildered guard was a tall, strong fellow about my age. I was unarmed, so made a vaguely threatening gesture towards him and he turned his attentions to me. He reached for his scimitar, and I knew I was finished if he succeeded, so I drew back my right fist and drove it as hard as I could into his face. He staggered, but didn't fall. I swung my foot up intending to catch him in the crotch, but he had evidently done a bit of brawling himself for he avoided the blow and seized my foot. Thrusting back, he sent me sprawling to the floor while he was left holding one of my curly-toed red slippers. Again he reached for that tremendous scimitar, so I threw myself upon him like a bob-cat, with desperation as my only plan. The impetus drove him back and we crashed through the partially closed doors. Rolling sideways to evade his grasp, I leapt to my feet.

A female voice gasped in terror behind me, and I heard Fourès yelling, "You little tart! You deceitful bitch! I'll teach you to betray me!" I turned partially and glimpsed the Lieutenant brandishing his sword-belt which he brought down with a tremendous

WHACK across the bed in front of him. Just in time a vision of loveliness sprang from the satin frills onto the floor! Forgetting my opponent, I stared open-mouthed. One must remember that this was my first ever glimpse of a pretty European lady wearing absolutely nothing. Not so innocent my opponent. I heard him scrambling to his feet behind me. At the same time the charming little creature darted in front of me, keeping me between her naked loveliness and her husband's avenging belt. I felt Fourès' strap sting across my shoulders as Bellilotte threw herself into my arms, the better to control my movements. Behind me the Mameluke roared as he leapt in front of Fourès. Next thing I knew, my opponent and I were back to back intervening most actively in the Fourès' domestic dispute -- the Mameluke fending off the blows of the sword belt, while I manfully shielded the plumply luscious Bellilotte whose brown tresses pressed against my chin as she clutched me. I reciprocated by clasping her shapely buttocks with both protective hands.

"Silence!" screeched the most commanding voice I have ever heard. Madame Fourès relaxed in my arms for a second, then pushed back. Indignation replaced fear on her round baby-face, and she cried out to the invisible presence behind me, "He's mad! He's a beast! I never want to see him again!"

"Silence!" screamed the strident, high-pitched voice again. We all turned to discover the source of such a screech. There stood a scrawny, pale young fellow wearing only a bleached nightshirt, a night-cap, and a baleful glare. Ridiculously skinny white legs protruded from beneath his nightie. Only his raging red visage and several straggles of long, unkempt hair brought colour to this scrawny phantom. I reacted the only way possible: I burst into laughter.

The spectre's glare settled upon me, stifling my ill-timed outburst. Struggling to restrain his anger, he hissed through clenched teeth, "You two have done well to protect Madame Fourès." Then taking a deep breath he exploded, "Now get out!"

Although the command was in French filtered through a strong Italian accent, there was no mistaking its intent. The Mameluke and I backed towards the door, bowing and scraping in unison. The eternal triangle remained glaring at one another. So ended my introduction to France's most famous citizen -- General Napoleon Bonaparte!

Outside the door, I stared at my opponent, noting the red welt I had given him under one eye. That convinced me to depart Sheik El-Bekri's palace without delay. As I edged away I saw relief spread over the Mameluke's face at having gotten out of the room, but in a flash he remembered the original state of affairs and turned to stare at me. I nodded and flashed what I fondly hoped was a friendly smile then set off down the corridor at a run. The marble floors made it difficult to keep my footing -- especially with one foot bare -- and I had no idea where to go. I soon found myself trapped in a courtyard with no escape. Seconds later the Mameluke in his leather-soled boots skidded into the moonlit space, wide-eyed astonishment on his boyish face, his arms flailing as he tried to remain upright. It was too much for me, and I burst into laughter. He glared at me. Then a flicker of a smile crossed his face for I must have been a pathetic sight standing there in my ridiculous red slipper with boyish clenched fists as my only weapons. Manfully he attempted to maintain his threatening glare, then he too burst into laughter. We remained there roaring hysterically. This laughter initiated one of the strangest friendships I have ever enjoyed -- or endured.

THE MacHUGH MEMOIRS ~ (1798 - 1801)

"Silence!"

A SECRET OF THE SPHINX

Roustam Raza was my new friend's name. He was nineteen years old, an Armenian-born Mameluke in the employ of the Sheik El-Bekri, in whose palace Bonaparte was now staying. We stood there laughing ourselves silly till common sense returned. By then to recommence fighting seemed even sillier. Although we could barely communicate in Arabic, by signs and smiles Roustam convinced me to follow him. I needed a friend, for sooner or later Fourès was certain to let slip that I was his "servant". I was certain the Lieutenant would be punished for the scene he had precipitated, and when my role was revealed I would share in his punishment. If it was ever discovered that I was British, the French would put me to death. The Egyptians, on the other hand, would probably welcome me as a potential liberator. Consequently, I followed Roustam Raza into the inner recesses of this vast house.

Eventually we arrived at a magnificently appointed apartment -- mostly silks and cushions. Despite my limited experience, I instantly recognized its femininity. Roustam called softly, for it was now well past midnight. In a moment we were surrounded by women! All were gorgeous -- and scantily dressed, but seemed to feel quite at ease in the presence of Roustam -- and even me. Roustam patted a few rounded buttocks and fondled an ample breast or two while he spoke in a wheedling tone with these friendly women. Then with a farewell gesture towards them and a vague salute towards me, he was off -- to stand guard once again at the door of the unfaithful Madame Fourès, I assumed.

What a delight it was to be fawned over lasciviously by all those buxom women! On the whole, I prefer slender, athletic ladies, while these charmers tended to be a bit broad of beam for my tastes. But I've never been a true esthete, and was willing to enjoy this overabundance without regret. After spending the rest of the night shared by two of these robust ladies I required the entire morning to sleep and regain my strength. It is due to my innate modesty, no doubt, that the next few days are a blur. The ladies kept me hidden away like a pet with special privileges. Although overwhelmed by lust, I am proud to claim that a tiny part of my mind clung to essentials -- that I must eventually get out of there and return to Mister Flemming. In the meantime, I did something which I persuaded myself justified my stay in this harem -- I began my study of Arabic.

CHAPTER 6
"Settling In"
January, 1799

Although my hideaway was one long frolic, it wasn't without danger. I discovered that I was guilty of defiling the wives of Sheik El-Bekri, an offense punishable by death. The Sheik had been made *nahib al ashraf*, or 'head of the sherifs' by Bonaparte, so I was trifling with the upper crust. It took only a short time to discover the reason for my popularity in the harem. Sheik El-Bekri was not an attentive husband. He seldom and only with great reluctance joined his ladies. Because the male attendants and guards were traditionally eunuchs, my appearance had been greeted with delight. Of course, the risk I ran was appalling, but I was young and foolish -- and randy. Although Roustam and the ladies took me to be a Frenchman, they liked me because they recognized that I was a fugitive from Bonaparte.

For his part, Bonaparte, the *Sultan Kebir*, inspired a mixture of emotions -- possibly the secret to his astoundingly successful career. People feared him, yet wanted to trust him; their hatred for the invader was mixed with admiration. One of El-Bekri's neglected wives once told me, "*Sultan Kebir* has come from God. He is His prophet. If Aboonapart so wished, he could simply point a finger at the Sphinx and make it crumble to the ground or sprout wings and fly into the heavens!" I was impressed and tried to recall the features of this man I had momentarily glimpsed in a ridiculous nightie and night-cap. I could not. It was a face that defied description; there were no special features to pin it in one's memory. Bonaparte's visage was simply a canvas upon which to display his emotions -- real or otherwise. I remember clearly that imperious glare -- and the instant effect it had produced upon the four of us. He was indeed a remarkable man. To justify my decision to settle in, I thought it would be a good idea to collect information about "Aboonapart" for Mister Flemming. By asking innocent questions to start bystanders talking among themselves I learned that Bonaparte had been very successful at putting down several "rebellions" against his rule, which he described as either the "French Republic of Egypt" or the "French Islamic Republic". His government was certainly no "republic" as dreamed of by revolutionaries. Using Sharia law and Islamic terms he disguised his tyranny from those not inclined to investigate. Thousands died in the rebellions, with even more dying in his retributions. The latter were carried out by the French soldiers bayoneting bound prisoners whose corpses or near-lifeless bodies were then thrown into the Nile late at night. Although the word soon spread that never before had the Nile's alligators gotten so fat and lazy, no accurate figure was ever released as to the number of his victims.

The great "Aboonapart" also inserted himself into Cairo's social scene. He issued orders that all inhabitants must wear cockades in the colours of the Revolution -- red, white, and blue. Sashes of the same colours were only to be worn by local celebrities -- chosen by Napoleon himself. These sashes were proof to common folk that their betters recognized the new regime. This was certainly not in any revolutionary tradition, but was very similar to the Royal customs of rank and privilege.

Bonaparte encouraged his senior officers to stage Balls and other social events -- partly to defuse the overwhelming boredom sweeping through the ranks of the French Army and partly to impress the local elite that the French were a cultured civilization. General Bonaparte even staged several Grand Balls himself. Indeed it was at one of these that he first spotted Madame Fourès, although he did contrive to encounter her again at one of his most dramatic displays. On December 1, 1798, Bonaparte had arranged for an unmanned balloon ascent as a propaganda coup. According to some of my sources it was during this event that he met the charming 'Bellilotte' and seduced her the same night. Unfortunately for him, his propaganda coup had not been as successful. The balloon caught fire and the gondola fell leaving the wind to scatter Bonaparte's printed proclamations to the locals.[12]

~

My new friend, Roustam Raza, often visited his master's wives, and was welcomed with open arms in every sense of the word. He was a genial and likeable fellow, but although we became friends, we seldom confided in one another. Charming, and even generous on occasion, above all else Roustam kept his eye upon his own advancement. Probably the events of his youth had shaped him so, for he had been kidnapped in his native Armenia at the age of seven. Eventually he had been purchased by one of the *bey's* in Cairo. A year and a half ago

-- while I had been hunting buffalo and stealing horses with the Blackfoot -- Roustam had accompanied his master on the *Hadj*, the sacred pilgrimage to Mecca. By the time they had started the return journey his master had learned of the French invasion of Egypt. He had turned aside from returning to Cairo and had gone instead to visit his old rival, Djezzar Pasha, in Acre. Djezzar had offered Roustam's master a cup of coffee, and his master had died half an hour later. I would later discover that this was, for old Ahmed Djezzar, the Pasha of Acre, a mild act of revenge. Roustam had wisely slipped away from Acre disguised as a peasant and made his way back to Cairo. Here he had been much impressed by the "handsome old grenadiers" he found in control, being particularly captivated by their big moustaches. Roustam Raza had turned himself into a Mameluke and taken service with Sheik El-Bekri, which led to our first meeting -- and would ultimately lead to fame, although short-lived, almost as great as that of Bonaparte himself.

During one of our meetings Roustam's hand found its way to my buttock. When I jumped away startled and ready to strike, he laughed easily and proceeded to fondle one of his master's many wives instead. When I mentioned the incident to one of the ladies I called 'Jasmine' (because I couldn't pronounce her name) she laughed and explained. "Roustam likes both of His creatures equally. You have noticed that our husband, Sheik El-Bekri, does not visit us. That is because he prefers boys, and for the present, Roustam is his favorite boy."

This at first surprised me. I had assumed that harems were simply for the sexual gratification of the nobility. As I became more familiar with Egyptian society I realized that harems were there to insure continuity of the family prestige and possessions.

It was vital for the *emir* or *bey* to produce a suitable male heir. A large collection of mates insured success, unless of course, the male's contribution was faulty or not forthcoming. I don't know how Sheik El-Bekri managed his hereditary matters, but that was the least of my problems. I did hear that most harems also contained a number of hidden cells where junior male descendents were kept in reserve until one of them succeeded to his father's position. That was the signal to quietly dispose of all the younger brothers, thus eliminating any legal rivals.

There in the lap of luxury I became somewhat careless, and often roamed through the neighboring apartments and gardens, for I have always been anxious to tempt Providence and sample forbidden fruit. One night as I was sneaking through a tiny garden, I spotted a truly beautiful girl. She sat alone in the moonlight, pensive and forlorn. Unlike the harem women I had been with, she was slender rather than voluptuous. I watched for several moments until bravado got the best of me. By this time the ladies had kitted me out to look like a posh Ali Baba. I suppose it was this rig plus my growing use of Arabic which gave me such foolish confidence. I stepped silently out of the shadow. She looked up, startled. "Do not fear," I said, reciting my rehearsed speech. "I mean you no harm. I have not seen you here before. Do you like this spot?" I began to realize the limitations of my Arabic, but by now it was too late.

"Who are you?" she demanded imperiously. "You have no right to be here. Servants are not allowed in my garden. Leave or I will have you flogged."

Noting my blue eyes, she exclaimed, "Why I believe you are French!" Then she switched to hesitant French, "Who are you? What are you doing here? If I tell Aboonapart, you will be punished."

I have never known a beautiful woman I could resist. This one was lovely! Darkly delicate and gracefully formed, with soft, glowing eyes -- despite the tough words -- she had won my heart, though that was the least of her intentions. I stood gaping for a moment before I managed a feeble response. "I am sorry to disturb you. I am new here and did not know."

"Then you do not know who I am?" she asked with disbelief.

"No," I confessed then added brashly, "but I would very much like to make your acquaintance, for I admire beauty."

She sized me up for a moment. "You are young and mean no harm I believe. What is your name, intruder?"

"I am Gaspard Genereaux at your service, mademoiselle." Although I had used the name assigned me by Mister Flemming, it was also a family name and came readily to my lips.

"I am Zenab, daughter of Sheik El-Bekri. You must have heard of me," she added with great bitterness.

"No, I have never before heard of 'Zenab'."

"Do not lie to me! There is no Frenchman who has not heard of me -- to my everlasting shame. It is I whom you call 'The General's Egyptian'! You Frenchmen have all leered and now laugh at the discarded concubine of 'the great Aboonapart'!" In the moonlight I could see the tears welling in her magical eyes.

"I have heard none of these things, but my eyes tell me that you are a woman of supreme beauty and worth -- a princess amongst peasants." She looked away, but did not dismiss me or raise the alarm, so I continued. "I see a lady of unbelievable grace -- beyond the power of rumor to harm -- and far above those who believe such viciousness."

She pulled herself together with an effort and smiled at me for the first time. It felt like the sun had just come out. "You know the right things to say, Caspar -- but nevertheless you are an intruder. Who are you, and why are you here?"

What could I answer? I was so love-struck that I fell back upon a version of the truth. "I am also one trapped by the snares of Fate -- Kismet. I am a friend of Lieutenant Fourès and came here to help him liberate his wife from the clutches of the Sultan Kebir."

"Ah! So you are the missing one! There is much talk about your visit. Aboonapart wishes it to be believed that no such visit occurred and has put about the story that the bitch Fourès had already divorced her husband. The truth is that after much shouting, her husband left declaring that he would divorce ***her***. There has been speculation about his unknown accomplice, but my father's favorite boy -- who stood guard that night outside the door of the yellow-haired adulteress -- claims the accomplice escaped onto the streets." Telling this story seemed to have cheered her up. "General Aboonapart has men searching for you everywhere. He does not want a witness to his embarrassment. He even expressed to my father his desire that the guard be murdered, but my father would not agree, for he is much in love with his Roustam. Where have you been hiding?" Seeing my hesitation, she assured me, "Do not worry. I have no desire to hurt one who has discomfited Aboonapart -- far from it."

THE MacHUGH MEMOIRS ～ (1798 - 1801)

"Roustam Raza, My Mameluke Brother"

"I have been staying with the wives of your father."

"Oh!" She looked shocked momentarily, then recovered. "We must not let you stay there. Those women will rob you of your virtue if you stay long enough. Although your Arabic needs improvement, you could pass as a Mameluke. You must become my bodyguard, Caspar."

∼

For the daughter of Sheik El-Bekri, it was an easy matter to appoint her own body-guard. Consequently, each night I stood outside her door till dawn. I didn't know from whom I was protecting her, but chose to believe it was General Bonaparte.

I became a Mameluke. My friend, Roustam Raza, found me a small room in the servant's quarters, and I no longer had to hide in the harem. He also taught me the military arts as practiced by this warrior class. I was impressed by the similarities with the Blackfoot, for both were fearless and both were expert riders. Formalities and extreme acts of bravery were also common traits. Both were fierce in attack, but likely to withdraw at top speed once it became evident that victory was not in the cards.

Mamelukes understood only three things -- horses, slaughter, and extortion. They had absolutely no grasp of modern warfare. The Mameluke, whether alone or part of a horde, always rode full tilt at the foe, firing off his pistols (as many as six) which he threw down to be retrieved by his *serradj*, a foot-soldier-cum-servant. Sometimes he would follow this barrage with a four-foot javelin. Finally, taking the reins in his teeth, he would draw a scimitar in either hand and throw himself headlong upon the enemy. A scimitar in the hand of a Mameluke was a fearsome thing. They had been trained from childhood to wield these antique curved swords till they could sever a head or an arm with a single blow. I trained diligently and became reasonably proficient using a scimitar in either hand, but never advanced to the point of handling two at a time. In return, I helped Roustam learn to speak French, as he showed a surprising interest in the subject.

One day while working out with scimitars under the guidance of my friend, another Mameluke sauntered up to look us over. He was one of the handsomest men I have ever seen. His perfect, regular features were set off by the dramatic upward sweep of his moustaches, but it was his large, expressive brown eyes that held my attention. "So I return from doing my master's bidding in Alexandria and find a new fellow has taken possession of my room." He said it in a matter-of-fact, almost jocular manner. Roustam continued his mock attack in silence while I parried. "Little blue-eyes," continued the stranger, "if you are taking lessons from Roustam Raza you are wasting your time. To learn the fine points of swordsmanship you must duel with a master swordsman." He leaned against the doorway, but I didn't respond, so he smiled insolently at my comrade and continued in a bored manner, "Roustam, old friend, you might for a short while be El Bekri's favorite, but you are certainly no master swordsman. If you teach this young man everything *you* know he will only end up being slain."

Roustam ceased his attack and leaned upon his scimitar. The new-comer waited for one of us to take the bait, and it was Roustam who responded -- looking at me. "Some say Omar of Myra is the greatest swordsman on earth." Then he turned towards the newcomer, put his hand inside his jacket, and added pleasantly, "*They* also say that he copulates with pigs and dogs in the street." He smiled challengingly. "Of course, wise men tell us that only half of what 'they' say is the truth. The trick is to know which half is which."

Omar of Myra swore and seized the hilt of his scimitar. Wrath had replaced the insolent smile on his dark face.

"Since the French have killed all the dogs and eaten all the pigs," continued Roustam with an expression of mock concentration, "it must be true that Omar of Myra *is* the greatest swordsman in the world." He laughed easily at his own jest. The newcomer turned abruptly and strode away with Roustam's eyes following him. My friend removed his hand from his jacket where I glimpsed the hilt of a knife. "Omar of Myra is just a swaggering Turk who thinks himself quite pretty. But he *is* a better swordsman than you -- and even me -- at least with steel," and he laughed again, "so you had best be careful, little brother."

Although almost a Mameluke myself, I have to admit I have never seen a more useless class of citizen. In luxury and ignorance they were borne upon the backs of the hard-working *fellaheen*, or peasants. The Mamelukes are a layer of parasites sandwiched between the Coptic tax-collectors and their Turkish masters. Fortunately, they did not multiply due to their predilection for sex with other men. Mamelukes served only themselves, squandering magnificently and bickering lethally. It was this relic of the Thirteenth Century which I represented each night as I stood guard wearing my huge *cahouk* (head-dress) and baggy satin *charoual* (trousers) outside Zenab's door. Although she preferred me to wear no jacket, I was loath to draw attention to my Sun-dance scars, so I normally wore the traditional *Yalek*. It was also a chance to indulge my love of bright colors.

During the ensuing weeks I fell more in love every night. To Zenab I was someone near her age who was intriguing and gave her life a purpose. She devoted herself to my "improvement" by teaching me the Arabic language and Muslim customs while at the same time indulging in her own small intrigue by simply keeping my secret. Of course we also shared a forbidden but unfulfilled sexual attraction.

In matters of sexual conduct, Egyptians were a puzzle to me. It was strange to see the higher-class ladies outside their homes moving about like little tents, totally draped in robes and veils so that not even their eyes were visible because women of standing -- and more importantly, their husbands or fathers -- would deem themselves disgraced if another man was to catch even a glimpse of them. Meanwhile at home their men-folk might be impregnating the female or sodomizing the male servants with no hint of censure. It was difficult to know what was acceptable and what was not in elite Egyptian society. However, everyone understood that the most unforgivable sin of all was for an Egyptian lady to have sexual contact with an infidel.

This fact brought home a mite of awareness to me, besotted and foolish as I was. At first I thought I might pretend to be what my sea-faring comrades called a "molly", but when I tried my version of the mincing walk I had seen performed once or thrice in London, Roustam burst into a fit of laughter. So I gave that up, and recalled Mister Flemming's Rule Number Three, "stick to things you know". The fact that Roustam was neither a eunuch nor a 'molly' suggested that Sheik El-Bekri was not as concerned with his harem as he should have been because here the normal requirement that all servants be either eunuchs or females was not strictly observed. After consultation with both Zenab and Roustam, I settled upon what I hoped was a solution to my dilemma -- I would become every inch a Mameluke. Hopefully, using that cover, I would not be considered a threat to the virtue of the female members of the household. Nevertheless, I lived in constant fear of being discovered and gelded -- or beheaded.

Zenab was a quiet and gentle creature, yet imperious by habit and upbringing -- quite unlike any girl I had known. Most nights she would order me to come into her room, and there she instructed me in the niceties of Arabic. We were both aware of the tremendous sexual charges that flashed between us, but for the time being played out the roles she had chosen. Zenab showed little fear of her father's vigilance, and eventually I understood why. The Sheik was normally befuddled by brandy and burgundy, both of which he consumed in great quantities. His pursuit of willing boys took up what remained of his evenings. Zenab, in her own quiet way, despised her failed father, the highest-ranking of all Egyptians. An upper class Muslim father was expected to be vigilant about the virtue of his wives and his daughters. How the man himself behaved was apparently not important. He retained his reputation for purity by insuring that his womenfolk remained inviolate. Yet Sheik El-Bekri had willingly allowed the seduction of the only Egyptian desired by General Bonaparte -- his own sixteen-year-old daughter. This "seduction" was more in the form of rape, for the girl had had no chance of repelling the General's advances once her father had turned a diplomatically blind eye to what was happening.

But Bonaparte had discarded Zenab once he had spotted Pauline Fourès, known best as "Bellilotte". An enthusiastic French mistress was more to his taste than a reluctant and inexperienced native girl who spoke almost no French. When I expressed my fears about Zenab's fate once the French left Egypt, she assured me, "The French will never leave Egypt. They are too powerful, and their homes are too far away. Aboonapart himself told me that he is planning to add to their numbers by importing slaves from inland, just like the sultan who created the Mamelukes did many many years ago. Besides, Aboonapart is to become a Muslim -- he told me so, himself -- and all will be well for me. Even the Imams will not dare to criticize me then. He will soon tire of his French woman, and will realize it is best for him to wed the daughter of the *nahib al ashraf*. My father has told me so many times."

Zenab's faith seemed touchingly naïve, but of course she had seen little of life and was shockingly ignorant of the world. As a general observation I must declare that I have never seen a more ignorant class than the women of the Levant. The lovely Zenab knew nothing about anything. Of simple mathematics or science, of animals or plants, of history or geography, she was blissfully unaware. She believed France to be in another universe, totally isolated from Egypt. She had no comprehension that Europe existed, let alone any understanding of the political events shaping her own world. To her, the universe was Cairo and the few miles beyond. She had not even been out of the city to see the greatest wonders of the world, the Pyramids and the Sphinx.

"Caspar', you are such a story-teller!" she would scoff, when I spoke of the world outside. Once, when I tried to tell her about the prairies and its enormous herds of buffalo, she remonstrated laughingly, "You mustn't think I am just a foolish peasant girl who will believe such enormous lies!" I confess that at first I dreamed of taking her there to share my life, but soon realized this was not something she would even consider. The thought of this gentle, passive creature facing up to Lark's Wing or any of the other Blackfoot girls I had known made me wince. No, if I were ever to win the daughter of El-Bekri, I would have to remain in Egypt.

That realization caused me to look about, to understand the people and the life into which I would have to be assimilated. Until now I had experienced life as a Scottish/French-Canadian city-boy followed by several years as a Blackfoot striving to achieve warrior status while enjoying the incredible independence of life on the Great Plains. Of course there had also been brief interludes of mob-rule in Paris and convict-life in London, but none of this had prepared me to understand what I saw around me in Egypt. At first I could see only the hypocrisy of those nearest to me, the inhabitants of Sheik El-Bekri's mansion. Later, when I ventured out of that restricted world, I began to understand that the most important fact of life in Egypt was their religion -- Islam. Most Egyptians lived by the "law of God", not "the law of man". Every facet of life was subject to the teachings of Mohammed as interpreted by various Imams. Of course, as in any religion, there were hypocrites who only pretended to live by the law and found numerous ways to twist or ignore its true meaning to their own advantage.

One precept that had suddenly been forced into prominence by the French invasion was Mohammed's dying edict that "infidels" should not be permitted to live in Islamic lands. Learning this, I understood why the French, despite their comparative leniency and generosity, were hated even more than the Turks and Mamelukes. This universal hatred meant that Napoleon's reputed dream of taking over the Turks' domain was doomed to failure. It also meant I would have to become a follower of the Prophet and enter a new way of life if I were to have even a ghost of a chance to make a life with Zenab.

Although deep in my heart I recognized that Zenab longed for General Bonaparte, I hid the knowledge from myself. I preferred to believe what she would have the world believe -- that she feared and despised the conqueror. Yet she still harbored hopes he would tire of Bellilotte and return to her. Zenab had not been told of General Bonaparte's wife, Josephine, waiting unfaithfully for him back in Paris -- or if she did know, preferred to ignore the fact.

Since I was a child I have never slept more than four hours a night, and this afforded me plenty of time to learn much that would be of inestimable value later. During the day I trained with Roustam, and worked hard to assimilate the ways of the Mameluke, because that path led to the best Egypt had to offer. I had seen nothing to suggest I would ever be able to return to my former life, so I enjoyed the many pleasures available to me. When I needed something I simply told Zenab, and it was provided. I lived in absolute comfort on the fringe of luxury, sampling the best in food, horses, and women.

One thing still nagged at my confidence -- Omar of Myra. Everyone else whom I ran into accepted me as a member of the household. Already several French soldiers had deserted and somehow found themselves places within the ranks of the Mamelukes, and no one, not even Bonaparte, seemed to be bothered about it. Omar of Myra, however, had lumped me together with Roustam, whom he obviously hated. I was new, youthful, and suspect -- the perfect target for bullying and mischief. Recalling his comment that "the new fellow" had taken possession of his room, I decided to go to him and proffer a small olive branch.

"Omar of Myra, others assigned me the room in which I now reside. I certainly do not wish to cause you any discomfort by claiming it. Please, in the spirit of friendship, accept the room as yours. I will find another."

I was amazed when he leapt to his feet struggling to master his rage. He hissed at me, "You cannot '*give*' me what I could easily take anytime I desire. Omar of Myra **takes** what he wants, little boy! You can *give* me nothing!" He took a step towards me, so angry that he almost sobbed, "T'is another who occupies my place!"

When I told Roustam Raza what had transpired he shook his head. "Little brother, you should never go meekly to a proud man. Now he despises you and believes you are afraid of him. Omar of Myra has killed others -- for much less. Watch your back, Caspar. You should have said nothing It is me whom he hates!"

"Why should he hate you?" I asked with astounding naïveté, forgetting the rule I had made for myself years before to stay out of other people's business.

Roustam gazed at me as though trying to read my thoughts. Finally he replied quietly, "I now occupy the place that once was his -- in the chamber of my master, Sheik El-Bekri."

I had been chafing to get out and around Cairo, for I had seen nothing during my arrival with Lieutenant Fourès. When I mentioned this desire to my friend, Roustam, he laughed and agreed to "make the necessary preparations" -- which should have warned me. On the chosen day we rode out the main gate preceded by a troop of men on foot, each armed with a big stick. Immediately, we were engulfed by a sea of humanity. And what humanity it was! Everyone seemed to tower above the French soldiers who roamed about unarmed and in small groups. I felt stifled, despite the efforts of Roustam's stick-wielding villains who literally beat our path through this jungle of humanity.

I later read that Bonaparte wrote as his first impressions, "Cairo, which has more than 300,000 inhabitants, has the world's ugliest rabble." For once I agree with the tyrant. Hundreds of blind men, crouching beggars, naked holy men, skinny children covered with running sores, milk-white Circassians, jet-black Nubians, and every shade in between mingled in a maze of narrow streets. The few wealthy, what the French refer to as *bourgeoisie,* were dressed in picturesque, vividly-colored costumes representing their many backgrounds. The vast majority of citizens, however, were draped in dirty rags. I had been shocked by my first encounter with Alexandrians, but they were high society compared to the citizens of Cairo.

The city was even less impressive than its inhabitants. The streets were a chaos of narrow, unpaved alleys, almost impassable during the day. Packs of wild dogs roam at will notwithstanding Bonaparte's efforts in poisoning several thousand in one night. Most shops looked like stables which had never been cleaned, while famous sites such as the El Azhar Mosque were infested with beggars and fanatical *fakirs*. Only when the human tide receded at sunset, could one see how truly filthy Cairo was -- with garbage everywhere, the occasional human body, and scores of dead dogs. These were my first impressions of Cairo.

Those impressions were to change dramatically once maturity accustomed me to the crowds, the noise, and the dirt. As has been the case ever since, I soon discovered countless examples of incredible beauty behind the ugly distractions. The Moorish art and architecture that dominate entire sections of Cairo are stunningly beautiful. If one looks upward, minarets and mosques of amazing elegance and grace reach into the bright desert sky. On a later excursion we rode up a small mountain to the Citadel which is itself magnificent, but even more impressive, in fact amazing, was the panorama of Cairo with its forest of minarets, domes and towers, beyond which stretches the endless desert broken only by the three pyramids, the Sphinx, and the Nile. To this day, Cairo stands out in my memory as the most amazing city I have ever visited -- a far cry from my initial impression.

That morning, however, as I rode through Cairo on my first outing, we happened upon an execution. I was not yet prepared to try my Arabic in any protracted speech, but I listened keenly to the crowd around me and heard an amazing story. Only four nights ago, three women of a household had been murdered by robbers. Although there was absolutely no evidence of any kind, public opinion held that the culprits had to be French soldiers. This very morning Bonaparte had returned from Suez and had been advised of the crime. Without making further inquiry -- let alone setting up a court-martial, or calling for evidence -- he had ordered the arrest of ten grenadiers of the

32eme Demi-brigade de Ligne. To the delight of the crowd, two were singled out for execution by firing squad. Roustam and I watched as they were marched out in front of the wall. They were not the magnificently-dressed grenadiers depicted in the prints I had seen. Gone were the elaborate blue coats and the huge bearskin hats. These two and their comrades in the firing squad which accompanied them were dressed in badly-dyed brown cotton uniforms with loose white pants slit up to the calf. On their shaggy heads they wore a remarkable new style of leather helmet with ear flaps tied up and out of the way.

The two condemned grenadiers were given an opportunity to say their piece so they asked for wine. "To our Commander-in-Chief," they toasted. "We hold no grudge against General Bonaparte. He has been led into this error by liars, for we are not killers of women. Now you will see how grenadiers of the 32nd face death!" Then they threw down their glasses and were shot by their comrades. The incident had a great impact upon me, for it was at that moment I first understood the men -- and the leader -- we were up against.

Several days later it became general knowledge that one of the family's servants had confessed to the Cairo police that he had committed the murders.

∼

It was on a night when Bonaparte was absent from Cairo that Zenab succumbed to temptation. She had called me in to instruct me on the basics of Muslim law, one of the few topics in which she had been educated. But we had held back the passions raging within both of us for long enough. This night, in the middle of her description of the *Hadj*, I kissed her, and by dawn when I resumed my post outside her door, although I remained ignorant of Mecca, I had explored Paradise. I stood in the corridor, outwardly impassive, but inwardly besotted.

I was hopelessly in love. We had no future, I realize that now. Maybe I even realized it then, although I had just turned seventeen. But it wasn't the future that concerned me. My life had been a series of misadventures into which I had stumbled one after another. Foresight and planning had seldom been part of my experience. Zenab and I were no longer discreet; we explored the physical aspects of love each night as soon as one of us decided it was safe for me to slip into her room. Our conversation was limited to the hours between dusk and dawn when we weren't tumbling about in her bed. It was, I now have to admit, a severe case of 'young love' rather than 'the love of one's life'. But at the time I believed it was the real thing, and I thought that she did too. I would willingly have died for her.

Then one morning the real world once more burst into my Paradise. I had just resumed my post as guard at Zenab's door when down the hall towards me tiptoed a servant of some sort. It was one of the eunuchs I had seen around the harem. He stopped beside me, looked nervously about, then whispered, "One awaits you beneath the head of the Sphinx at midday. Wear this sash. You will recognize another who wears its twin." Before I had collected my wits -- which had been abandoned amongst the tumbled silks of Zenab's bed -- he had padded off down the hall and disappeared. In my hand I found a multicolored satin sash.

The sight stunned me! I had emerged from the tangled streets of Cairo on my prancing stallion, and there on the flat horizon were perched three Pyramids, for all the world like the Sweet Pine Hills of Blackfoot country. These three were regular in form, and lacked the wildness of my three special buttes. Nevertheless, the sun-drenched panorama brought me back to my Blackfoot heritage with a jolt. Reaching inside my *yalek,* I felt my Medicine Bundle against my chest. Black War-Bonnet was once again in touch with his Vision.

In the "Old City", which stained the east bank like a scum of effluent, I enquired about the Sphinx and learned that it was regularly visited by the French invaders, and even by locals. Consequently, it was not difficult to be ferried across the Nile to the Isle of Rhoda, then to take the bridge to Giza. At last I emerged from the Giza Gate onto the Libyan desert which stretched westward for hundreds of miles. Immediately before me towered the greatest wonders of the world. Three golden-tawny Pyramids dominated the landscape, but my goal was the Sphinx. This famous relic of Egypt's mysterious past was largely buried in the drifting sands. All that was visible was the creature's back and great stone head rearing out of the sand. Behind -- the world's most dramatic man-made backdrop -- towered one of the Pyramids. I stopped to admire the amazing panorama before cantering down the slope to dismount below the enigmatic stone face gazing across the ages.

There were few people around -- half a dozen French soldiers and as many Egyptians. None wore a multicoloured sash like mine. Swaggering like a true Mameluke, I played the part of a tourist well, for I was very much drawn to the Sphinx. Possibly it was the re-awakening of my Blackfoot heritage that morning, but something spiritual drew me to the mysterious, impassive face. It had not survived the eons without scars. A large piece of the nose was missing; otherwise it would have been an attractive sculpture. I later heard it said that Bonaparte's cannons had blasted it off during the so-called "Battle of the Pyramids", but I saw no evidence that the damage had been that recent. The battle Bonaparte (always an excellent self-propagandist) had so grandiloquently named had actually been fought many miles away.

Wilting under the blazing mid-day sun, the other tourists soon headed back towards Giza, leaving me alone beneath the colossal chin. Gingerly I reached out to touch the ancient stone. A shock went through me, and I slipped the hand into my *yalek* to hold my Medicine Bundle. For a moment I saw my mother's face as I stared upward at the giant head. She was trying to speak to me -- a warning -- but I could not hear her words. The nickering of a horse made me whirl about, and there sat my Blackfoot brother, Hunts With The Owls, impassive at first -- then a smile broke his dark face. I blinked against the sun's fierce glare, and the image changed ever so slightly until I recognized Roustam Raza!

"My little Mameluke brother!" he said with his broad smile. "So it is **you**!"

"Roustam! Allah go with you," I exclaimed. "I've heard so much about these wonders of yours that I came out to see them for myself. They are magnificent, aren't they? But why are you out here? You must have visited this place many times."

"I was sent to discover one who wears the sash you have around your waist. I am surprised to find **you**, however. I did not for a moment suspect that my brother was a British spy."

He urged his stallion several paces closer to me, and the smile left his face. "Are you prepared to die, Caspar?"

"Any time," I replied in confusion, seizing the hilt of my scimitar.

"Do not draw your sword on me, little blue-eyes, for then you will surely die," he warned. "Those who do such work must be prepared for an agonizing death if caught by the French infidels." Then his face broke into his normal dazzling smile and he added, "We work for the same people, little brother." He reached into a pouch on his saddle and pulled out a few inches of sash identical to the one I wore. "I bear a message for the wearer of the other sash," and he handed me a small, thin envelope, badly wrinkled. I opened it and read the following in Arabic:

"10 February
Today Bonaparte left Cairo for Katia. He plans to seize El Arish, then lead 13,000 infidels into Syria.[13] *May Allah curse them. Destroy this letter."*

I reread it several times, then tore the note into a hundred tiny fragments which I tossed to the desert wind. As I did so, I looked up at Roustam, puzzled. He grinned and announced, "I am to take you to the coast. There is no time to waste. We leave now. Mount up, little brother."

"Yes, but first I must speak with someone at the house of Sheik El-Bekri. Then I will be ready. It will take only a short time," I protested, for I could not leave my darling Zenab without the traditional romantic farewell.

"There is no time for such things," he said with a leer that suggested he knew all about my nightly 'duties'. "We go to do man's work -- a task truly worthy of a Mameluke. Mount up. We leave now. You must reach the big boat before it sails away from the Delta."

I mounted without further protest, but something held me there inches from the giant stone enigma. I could not stop myself from placing my hand on its wounded face. This time there was no shock and its warmth soothed me. We exchanged a long intimate gaze before I murmured my farewell. The Sphinx, I realized, was not finished with me.

On the ridge behind which it slipped out of sight for the last time I reigned in and paused as Roustam rode off. The Sphinx's impassive gaze was directed straight at me, boring into my heart! It was warning me! Suddenly above the great stone face appeared the mystical image of the Great White Owl -- my Spirit Guide! Ever so reluctantly, I turned away from my two guides to gallop off and rejoin my companion. Private MacHugh had been recalled to duty.

CHAPTER 7
"Recalled to Duty"
(Jaffa, Syria; 3 - 9 March)

"You will wait at a *caravanserai* on the Bethlehem road, MacHugh. There you will meet one of my couriers. He's an Egyptian, traveling with Bonaparte's army. We know the French have captured El Arish, and have moved on towards Er Ramle. Our man will be wearing a sash like yours. He will have a report for you to bring to me." Several days ago James R. Flemming had given me these orders, and now it was March 3rd, and still I waited for the courier to appear. I remember how surprised Flemming had been when I had turned up, and how concerned he had become when he calculated that there had been a delay of two days between the writing of the letter and its delivery to me. "She is normally so efficient," he had commented. He was even more puzzled when he learned that no one had approached me with the contact words, 'The nightingale's song is so sweet".

Flemming had set me ashore at Jaffa, a large Turkish garrison on the Mediterranean coast. Garbed as a Mameluke again, I had fallen easily back into the life of arrogant indolence. Using the generous funds with which he supplied his couriers, I had purchased a good mount and had ridden southwest a considerable distance to wait at the appointed meeting spot, a *caravanserai* on the road from Bethlehem and Er Ramle. I wiled away the time dreaming of Zenab, and wondering if she missed me. Back in Cairo the weather had been warm and dry, but since entering Syria I had discovered that late winter meant constant rain and cold winds. It was not how I had pictured the Biblical world. Fortunately, the days had been growing balmier, and at Jaffa on the coast spring had arrived.

Three days had passed since the expected meeting with Flemming's agent. On the first day swarms of Moslem civilians had passed by in panic, heading to Jaffa. They were the residents of Er Ramle, fleeing before Bonaparte's army. The torrent had slowed to a trickle by nightfall. I had been sitting here for almost two days now watching the empty road from the shelter of the vast, empty *caravanserai*. At last I saw a lone rider approaching along the road from Er Ramle. In the vacant central courtyard I met him, confident he would be the long-awaited courier. He was not. Instead, he was one of the Greek soldiers employed by the Turks. Strangely, he was clad in the remnants of a French uniform. To avoid suspicion I told him I was waiting for my brother, but I need not have bothered. He was much more interested in telling me his story while he rested his horse.

"Aboonapart is a Devil! We surrendered to him after resisting with all our might at El Arish. The town's innocent inhabitants had already been put to the sword by the accursed French, so we were encouraged to resist even more bravely. Eventually we agreed to surrender only after that Devil promised us that if we swore not to serve against him he would let us march away to Baghdad. Instead, after we had marched out of the fort we were forced to join his forces or die. Of course, we joined, but we do not owe loyalty to such a perfidious devil, and many of us have slipped away. I intend to join Abdullah Aga in Jaffa. If you wish to avoid capture and probably death, you had best mount up and join me," he advised. "The French are not far behind me. They intend to take Jaffa, but with Allah's grace we will stop them there." At this news the remaining employees of the *caravanserai* disappeared, presumably onto the road towards Jaffa. "You have missed your brother, I am afraid, for there is no one between me and the invaders. He has been killed either by the French or by the great sickness which is coming with them."

Upon first sighting the Greek I had prepared myself for his arrival by recalling the 'give-aways' taught me by Flemming. The Greek had employed none of them so I took his story at face-value. As I mounted, he prepared to head towards Jaffa, while I rode off in the opposite direction, up the road towards the advancing French. Soon I entered a hamlet clustered along the main road. It seemed deserted but gave me an ominous feeling so I slowed my mare to a walk. All at once I came abreast of several cavalry horses, heads drooping from exhaustion, standing in the mud between two buildings.

At that moment two Frenchmen appeared in a doorway. Their arms were full of loot. I had no quarrel with them, so I reined-up and studied them. Clad in torn remnants of uniforms they looked like scare-crows. Were these the dreaded soldiers of the Republic? My own experience of fighting-men was limited to the Blackfoot and their enemies. These two looked nowhere near as fierce.

BLAM! A musket roared behind me, and as I turned the shot whined above me. More looters blocked my retreat! Although burdened with loot, they began reaching for their weapons. One dangerous-looking individual was taking careful aim at me and I feared my time had come. Suddenly from between two houses a bolt of colour shot out behind him -- a horse and a rider in French rags. The rider's sword swept down upon the marksman, cleaving his skull.

At last my mind began to function. In one motion I put my spurs to the mare and drew my own scimitar. Like Kills Twice would, I bent low over the saddle and charged straight into the cluster of looters. Raising in the stirrups, I swept the broad blade down at a hand as it grasped my bridle. The blade jarred as it met bone. Blood sprayed over us as the Frenchman screamed and fell away from his hand which clutched the bridle for an instant before it too fell. I was vaguely aware of the other rider barrelling along abreast of me. The road was open before us, and I galloped out of the hamlet whooping a Blackfoot war-cry. It wasn't Caspar the Mameluke who rode back towards Jaffa that afternoon: Black War-Bonnet made the journey.

When we were out of gunshot range I took the opportunity to look at my rescuer. It was the Greek from the *caravanserai.* "You saved my life. Thank you, friend," I said as sincerely as I have ever spoken.

He grinned in reply before explaining, "I couldn't let you waste your life upon that garbage. You are just too young to understand what they are like. Looting is how they spend almost all their time -- that, and rape, and murder," he assured me. "Never has this land seen such voracious looters. Let us pray that the great plague they carry with them will wipe them from the face of the earth before they steal everything and murder everyone. Now let us get moving. My lovely wife and our little *Janissarie* await me in Jaffa."

~

My rescuer's name was Leonidas Kapadopolis. While we rode he continually divested himself of anything connecting him to the French invaders. It wasn't long until he was stripped to the waist. One could not help but admire his physique and the unique tattoo which covered most of his right arm. "That is the badge of my *Ortah*, what the French call a 'Regiment'," he explained. "We Janissaries wear our badges for life. Some *Ortah's* even have their badges tattooed on their foreheads." All afternoon he regaled me with stories of his career as a Janissary. Although his adventures had brought mostly negative results so far, his sense of humour had not been quashed, so we spent a lot of time laughing on the road to Jaffa.

The only thing that occupied my new-found friend's heart even more than his *ortah* was his small family. The vagaries of war caused by the French invasion had separated Leonidas from his small son and "the most beautiful woman in the vast Ottoman Empire". Once we arrived in Jaffa, he promised, I would be the first guest they would entertain -- but only after a delay to enable him to fulfill his "duties as a husband" as he delicately described it.

We approached Jaffa in a relaxed and a jovial mood. The ancient city had been built upon a conical hill. With the blue Mediterranean sparkling in the background and its orange groves laden with ripe fruit in the foreground I imagined the Garden of Eden. It seemed appropriate that the warm air bore the fragrance of millions of blossoms. Half way up the hill an ancient wall stretched between a pair of towers. Behind the wall rows of houses could be seen rising towards the summit, much like seats in an amphitheatre. Near the center was the citadel. About the crest large groves of almond and citrus trees were clustered, fitting my concept of a Biblical city. It was difficult to believe that anyone, let alone the rag-clad gang of looters I had encountered, would even contemplate attacking such a wonderful place.

With a flicker of clarity, I remembered that these scare-crows were the same *canaille* who had murdered my mother in a convent. Soon I would meet them in a real battle. Wouldn't it be great if that cowardly bully, "Moses Roses" was one of them! Surely my Spirit Guide would pick him out for me. Then I would avenge the first great injustice of my life! Then I could return to Sun Bull's lodge, my Sun Dance Vision fulfilled. But first, of course, I must sail back to Egypt and claim Zenab for my own. Such was the foolishness which occupied my mind as I re-entered Jaffa and the real world beside my newest friend, Leonidas Kapadopolis.

～

I was dismayed to discover that during my long futile wait for Flemming's agent the British gunboat had left Jaffa, and I now had no choice but to remain in the fortress under siege. The garrison was made up of about 5,000 Turkish soldiers, mostly the famous Janissaries. These troops wore fantastic uniforms straight out of "The Arabian Nights". Like Leonidas, some were bare-chested, others were swathed in layers of robes, but each wore an outlandish headdress unique to the *Ortah* to which he belonged. As with the Mamelukes, individual combat was their forte, and each man was responsible for providing his own weapons. Control of the Janissaries by their officers seemed tenuous -- no better than the discipline of a Blackfoot war party.

Because I belonged to no military unit and had no officer except the Turkish commandant himself, the aged Abdullah Aga, I drifted from unit to unit, eventually becoming part of a small collection of Mamelukes who had been trapped in the citadel. True to their traditions, I continued to roam when and where I wanted. Descending into the old city on evenings I enjoyed the hospitality of the locals, and after a day or two that of my new friend, Leonidas. His charming wife and three-year-old boy had brought an even brighter gleam to his sparkling eyes. 'So this is what married life is like for a soldier', I thought as I sat at the table with them, bouncing the lively little boy on my knee. 'It must be precarious, what with wars, long separations, and the constant threat of death to deal with.' Although Kismet blessed me with only one evening shared with the Kapadopolis family, I found a new understanding of the complications of the soldier's life -- and a rare glimpse of true love shared.

As always, I tried to soak up all available information about the land, the cultures, and the languages of the locals. But mine was destined to be a short education, for within three or four days the French appeared before the walls of Jaffa, and I realized that a savage attack was imminent. My memories of those few days are vivid, but unconnected, for I was still but a youth, and the assault on Jaffa was not only my first siege but my first European-style battle. It was early afternoon of March 7th when Bonaparte commenced his onslaught -- immediately after his surrender terms had been rejected by the garrison. The tall swaggering Janissaries were confident, as was I, that we could withstand any and all assaults by these small infidels in their ragged peasant clothes. But we were to be stunned by the rapid course of events! Even without siege artillery, the French managed to breach the ancient city wall below us within hours!

Through the breach they swarmed, like ragged madmen. To me, in their faded red and brown jackets they looked more like British infantry than French. Not far behind the first wave a group of officers were clustered. I recognized Bonaparte amongst them. The courage of the attackers was amazing! Although the Janissaries fought bravely, they were driven back from the breach, through which the French continued to pour. What happened next defies my powers of description. It was the first time I had ever seen that hideous phenomenon -- the madness that overtakes besiegers when they enter a city. The French went berserk! Everyone they came across was butchered with bayonets or clubbed muskets. Even after the Turkish soldiers surrendered by throwing down their weapons, the slaughter went on unabated. Christians were killed as readily as Muslims. Babies, women -- none were spared. The sickening scene carried on till the torn corpses of 2,000 soldiers and even more of the defenceless townspeople lay in hideous piles. Looting continued all through the night amid cheering, drunken laughter, and shrieks of pain and anguish. By dawn, only Frenchmen were left alive in the city below.

That night in the citadel was agonizing. There were between 2,000 and 3,000 troops enclosed in the small area. Most had lost loved-ones in the massacre, and we knew what to expect when it came our turn in the morning. It would cost the

French to break into the citadel, claimed some, for we would sell our lives dearly. But not everyone was willing to do so. Many thought they held trump cards and could negotiate an honorable surrender with Bonaparte.

I retained a strange serenity throughout all this -- at least as far as my own safety was concerned. My Sun Dance Vision had promised many things which had not yet befallen me. Therefore, I was confident I would live through all these horrors to return to the Great Plains. I was also young and inexperienced, so found myself almost a spectator in the events that followed, although logic should have told me that my life was at stake.

First came a meeting that shattered my illusions. While heading to take a position on the wall overlooking the ruins of the city, I encountered Leonidas. He looked forlorn and desperate, a man whose dreams had been destroyed. Holding his hand was his small son. "Oh, Caspar my friend," he moaned, "my life has been destroyed. My beautiful wife has been murdered -- raped over and over then murdered by those devils. She was so loving, so innocent. Now she is gone, ruined and destroyed, just like my life and my little boy's. I must somehow carry on until I achieve vengeance."

I tried to soothe him, but without success, partly because of his desperation and partly because of my inexperience in consoling others. At the same moment the leader of the Mamelukes I had joined ordered me to rush to the wall to repel an expected attack. Consequently I saw no more of Leonidas and his son for several days.

Sometime on the morning of March 8, two young French aides-de-camp -- conspicuous because of their fine uniforms and bright sashes -- appeared outside the citadel. Some of the Janissaries called to them from the windows of our fortress. "We are prepared to surrender!" they shouted. "We will march out, if you promise us safe conduct! Swear not to treat us as you have the inhabitants, and we will yield to you!"

The two young aides-de-camp conferred for a moment, then one, a Captain, replied. "I guarantee your safety if you lay down your arms and march out immediately. Come out now, and your lives shall be spared. You have my word as a French officer."

We did so, although many of us had doubts. But the optimism of the majority ruled, so out we marched and lay down our weapons. These two youngsters almost as young as me, one of whom I later discovered was Bonaparte's step-son, Eugène de Beauharnais, proudly led a column of nearly 3,000 captives from the surrendered citadel into Bonaparte's camp. The French immediately set about separating our party into three groups -- Turks, Egyptians, and Moroccans. Being to all intents and purposes a Mameluke, I joined the Egyptians. There was a feeling of optimism and much speculation as to our future. The madness of yesterday's assault had ended, some assured us; the French bloodlust had been satiated, and cooler heads ruled now. Once again 'Aboonapart' was in control, the optimists pointed out. We would probably each be given a bowl of rice and sent on our way after giving our parole not to fight against him again. So passed a long, hungry day filled with such speculation. The night was cold and comfortless as we continued to sit in front of Bonaparte's tent. No water or food was permitted us, and I could hear the sobbing of some of the children who had joined their fathers in the citadel after escaping yesterday's massacre.

The next morning, March 9, we watched as the Moroccans were marched away towards the beach, escorted by two battalions of French veterans. Soon we heard firing accompanied by shrieks and cries for mercy. None was given. The Moroccans tried to escape by running into the surf and swimming away from shore, but they were picked off by the French, some of whom were laughing while most were simply business-like. Many of the Moroccans swam to a cluster of rocks well out in the harbor, so French officers ordered several boatloads of soldiers to row out and murder the helpless survivors. By noon all 800 of our Moroccan comrades had been slaughtered.

Incredibly, some vague optimism remained amongst the rest of us. "Aboonaparte will be satisfied now. He has proven how powerful and ruthless he can be. He will now show what a merciful and just ruler he is by freeing the rest of us," went the theory. "After all, there are all those children with the Turkish artillerymen. He could hardly slaughter their fathers and leave the children orphans! Aboonaparte is a mighty warrior; he makes war only upon men."

Such desperate optimism appeared justified when we saw Jaffa's aged Commandant, Abdullah Aga, dragged, screaming, to meet the conqueror. The wretched old man threw himself at Bonaparte's feet and begged for mercy. His grovelling and abject terror disgusted his men, and the grossness of it convinced me never to beg for anything. Bonaparte first made a show of sternness, then granted the despicable old duffer his mercy. He promised Abdullah Aga his life, and the old tyrant was taken away sobbing tears of joy.

By the next morning, after two days without food or water, everyone was apathetic. The Turks, numbering 1,200 plus their children, were then marched away. As they shambled by I glimpsed Leonidas, who nodded sadly to me. His three-year-old son whose name I don't even remember clutched his hand as they were herded forward in the throng, but he waved hesitantly at me and broke into a wan smile. Although the dry lump in my throat prevented me from crying out, my eyes were able to produce tears of true anguish at this final encounter. All our illusions had vanished. The prisoners marched in silence -- except for the weeping of the children -- into a square formed by French troops. Moments later firing commenced, but soon ceased, for to save ammunition the French switched to their bayonets, "the weapon of the brave", as Bonaparte liked to call it. Hundreds of children clinging to their fathers were bayoneted by Bonaparte's brave heroes. At one point the Turks threw up barricades made of the bodies of their murdered comrades and, although unarmed, tried to fight off the French. It was to no avail. On the orders of General Bonaparte every man and child was murdered there on the beach.[14] If ever I learned how to hate, it was there on the beach at Jaffa.

By now my faith in my Vision had faded. I realized that I was about to stare death in the face. To die facing the foe did not faze me. Many times on the Prairies I had faced an enemy, and I knew the stench of death. But to have no chance to defend myself, to be slaughtered like one might squash a bug, was so humiliating and degrading that I almost yielded to tears of frustration.

Zenab also occupied my thoughts: I would never see her again, never touch her, never make love to her. She would never know why I had disappeared from her side. The thought of Bonaparte ravishing her again made me rage inwardly. To think that the first time I encountered him he had stood only a few feet away! Over and over across the years I have reviewed that preposterous scene: Bonaparte, raging away in his nightie, had appeared so ridiculous. Although I was unarmed I know I could have killed him. Instead, I had laughed at the ludicrous scarecrow. I could have sent him to Hell and saved all of those lives. Yet that would have been the cold-blooded murder of an unarmed man. I would have been no better than Bonaparte, and my conscience would have tormented me the rest of my life -- short as that might have been. But think of the millions of lives I would have saved! Ever since that night on the beach at Jaffa I have lived with that one regret. I have brought death to so many yet missed my chance to save so many by destroying the Devil incarnate, Napoleon Bonaparte.

That evening French soldiers brandishing their weapons, entered our huddled mass. They beat a path to where our small party of Mamelukes had gathered. I clambered to my feet to sell my life dearly, but was seized and man-handled back through the cowed prisoners. Four of us were thrown down in front of Bonaparte's tent. If I was going to die, it would not be cowering on the ground. My mother had died at the hands of these vermin, but she had died valiantly. Certainly her son was not going to do worse. I struggled to my feet and glared about me. "Try me one at a time, *canaille*," I croaked in French through my parched throat, " -- if you dare!"

"Face me, fellow!" came the familiar strident voice from behind.

I turned unsteadily, ready to attack. It was the Devil himself. Bonaparte's thin face, though still youthful, seemed almost haggard; his hair was long and straggly. How could such a man command such an army? He looked us over without passion. The others had also stood to face him. Two were grizzled old veterans; the third was a boy little more than ten years old, but every inch a Mameluke. We stood proudly, disdainfully, before the butcher.

"I recognize you from somewhere," said the General casually to me in his high-pitched voice. "You just spoke French. Can you carry on a conversation in French?"

"But of course," I gasped boldly. My throat was so dry I could barely be heard. I wanted to kill him, but had no means at my disposal.

"Have our paths crossed before?" he hissed through his teeth.[15]

"Not that *I* recollect," I whispered in French. "I don't remember *every* piece of shit I meet," I continued, hoping to irritate him. 'Just come two steps closer,' I prayed. Then the sudden mental image of this cold devil violating my beloved Zenab made me lunge forward, but in my weakened state I was easily pulled back by the guards.

"This young fellow seems to speak French -- and has spirit. Take these others back," he ordered, and my three comrades were escorted back to the starving mass of captives huddled in the growing darkness. "I have need of a messenger. I have written to my adversary, Ahmed Djezzar, Pasha of Acre. You will take my letter to him. You will help me save many lives." Then as though I was not even there, he looked around his circle of admiring officers and announced, "Here is what I have written to the old butcher: 'Since God gives me Victory, I wish to follow His example and be merciful and compassionate, not only towards the people, but also towards their rulers'." He laughed. Some of his officers took the cue and chuckled, while most wore only bleak expressions on their faces. "'I have treated with generosity those of your troops who placed themselves at my discretion.'" Here many stricken and guilty glances were exchanged, but Bonaparte never even caught his breath. "'I have been severe towards those who have violated the rights of war. I shall march in a few days to Acre.' Ha! This should help the senile old duffer make up his mind, **and** smooth our way into Acre and beyond -- all the way to Constantinople -- for there is nothing between Acre and the Bosporus to stop a company of good cavalry. When Acre falls, the Levant -- **all** of Asia, in fact -- is **mine**!"

I was led into the General's tent and handed over to Captain de Beauharnais, Bonaparte's step-son. "Sit," he said gently in Arabic, pointing to a chair. I sat and looked wearily about me, so starved and thirsty that I could no longer even swagger. "We should get him something to eat and drink, Crozier," he said to another young officer sitting dejectedly in the corner. "It's the least we can do for the poor bastard."

The other looked up, his face absolutely ashen with anguish. "Yes, it is the **only** thing we can do for him -- for all of them. One flagon of wine and a crust of bread to atone for the betrayal of thousands of lives. Eugène, I shall see their faces and hear their cries for the rest of my life! Yet I was betrayed as much as they were! Your step-father -- forgive me, Eugène -- our General -- has murdered me every bit as surely as he has murdered those prisoners -- men to whom I pledged my word of honor as a French officer!" I suddenly recognized the young Captain who had taken our surrender three days ago. Hatred surged through me, and I made to rise from my chair, and would have seized his throat had I possessed the strength. "See! This poor fellow recognizes me as his betrayer, the new Judas. Give him your sword, Eugène. Let him give both of us a small measure of justice, for only Death can succor me."

I collapsed into my chair. Captain Crozier was suffering far more than me, and his anguish would only end when his life was snuffed out. His honor had been betrayed by the man whom he served. I felt compassion for him. Young de Beauharnais gave me a few morsels to wolf down along with a flagon of wine, which went straight to my head. In a few moments I was taken outside where I was mounted on a waiting pony and dispatched towards Acre. My Vision had been right: I would live to see and do much more.

CHAPTER 8
"The Two Butchers"
(Acre, Syria; 14 March -- 9 May, 1799)

I arrived at the gates of Acre on 14 March, bearing Bonaparte's letter to the Commandant, Ahmed Djezzar Pasha, "Ahmed The Butcher" -- for that is what his name meant when translated into English. Although I had heard of his tyrannical rages and his habit of punishing messengers for messages he did not like, I was not prepared for my reception. I was thrown to the floor and my arms pulled behind my back by two burly guards in gaudy costumes.

"I'm British! I'm a courier for Sir Sidney Smith," I protested in Arabic. "Bonaparte has just slaughtered the garrison of Jaffa after they surrendered. Except for the Commandant, I am the only survivor!"[16] I continued desperately, for they had stretched me out ready to behead me. "Bonaparte spared me only to carry this letter to you, most noble Seraskier. I wish only to serve you and my country, Britain, by fighting the French infidels!"

He tore the letter from my hand. When it had been read to him, Djezzar became even angrier, and his long white beard shook with rage. I remembered stories told about his treatment of enemies -- even his most loyal allies. He had had their noses or ears cut off, eyes gouged out, and had even ordered some to be shod like horses.

"Off with his head!" he roared, pointing at me. "But first, cut off his hands and genitals!" When no one jumped forward immediately he shrieked, "Djezzar has commanded! Djezzar's commands will be carried out immediately!"

"If I might intervene, Your Excellency," came a calm voice in English with a French accent. "This young man is one of our most trusted agents -- **and** my personal friend." While someone translated this to the enraged old tyrant, I twisted around to catch sight of Count Phélippeaux in a Turkish Colonel's uniform complete with a brace of pistols. Unruffled, he continued, "I would be much distressed, if anything were to happen to my friend, as would his protector, Sir Sidney." Phélippeaux stepped between Djezzar and me, his charming smile assuring "the Butcher" that he had not taken offence at my predicament.

"Fools!" bellowed the old tyrant to his guards, "Djezzar orders you to release this fellow! **Now**!"

I sprang to my feet, keeping well behind Phélippeaux. In a moment we were out of the 'The Butcher's' presence. "We had best keep you somewhere safe, out of sight of the Pasha for a few days, MacHugh. My, you do have a penchant for getting into scrapes, *n'est ce pas?*"

~

I have read somewhere that histories of sieges are always tedious. I would have to agree -- both tedious and confusing. Day follows monotonous day; events repeat themselves endlessly. First the enemy appears before the walls and commences trenches which advance slowly. When close enough he attempts to batter a breach in the wall and dig mines to blast away the foundations. Both besieger and besieged endure constant cannonading while they wonder if supplies will hold out and speculate about whose reinforcements will arrive first. Inevitably, the inhabitants make sorties to try to destroy a trench or blow up a mine that threatens their wall. The besiegers counter-attack, and the besieged fall back inside the wall to resume their lives -- that of prisoners being slowly battered into unconsciousness. Finally those in command of the besiegers decide the breach is big enough, and order an assault. Both sides then fight madly to retain possession of that shattered section of the wall. If the attack fails, the besiegers try again, and again, and again -- and the defenders rouse themselves from their nightmare-world to throw their assailants back -- again and again and again. It finally ends when the attackers either break through and slaughter the inhabitants or give up and march away. By then, everyone has lost all track of time. The Siege of Acre lasted sixty-one days -- so I have read somewhere.

I spent nearly all my time with Colonel Phélippeaux as an exceptionally lowly aide-de-camp. The Colonel was an inspiring and reassuring figure appearing wherever and whenever a decision had to be made. Totally calm and without pretence he would suddenly arrive at the moment of crisis resplendent in his Turkish uniform augmented with a brace of business-like pistols. But he was much more than an inspiring hard-driving leader. He was a great soldier and an even greater man. Later Sir Sidney Smith was given most of the credit for events at Acre, and he was certainly deserving, but it was his friend, Colonel Count Louis-Edmond Phélippeaux, who was the real hero of Acre and maybe even the saviour of Syria.

Phélippeaux became my *beau ideal*. He was an exile from his homeland, fighting his own countrymen on a matter of principle, attempting to right the wrongs they had inflicted upon the world. To him it was not a matter of getting even; Phélippeaux simply believed he must do his bit to restore a small measure of honour to France. He appeared serenely confident and devoid of all fear, and that was no bluff. But the Count was also capable of warm friendship. Colonel Louis-Edmond Phélippeaux was like a much-needed big brother to me.

I attended the Colonel on his constant rounds of the fortifications -- if such they can be called, for Acre's wall had been built by the Crusaders as their last bastion in Syria five centuries before. Now the long wall was crumbling. Fortunately, the infidel Crusaders had built thick walls, a fact that might yet prevent Bonaparte's modern infidels from retaking Acre -- now, ironically, the last bastion of the "true believers" in Syria. The city sits at the end of a narrow peninsula, so there was only the one L-shaped wall. The real strength of the place was the sea that surrounded us on three sides. Our few ships were able to anchor on either side of the narrow neck of land and pound away at the besiegers. Because Phélippeaux was an expert on artillery and fortifications, Sir Sidney Smith had sent him as an advisor to Djezzar. The old butcher had been preparing to withdraw from the city and turn the place over to the French. Although Djezzar was a real fighter, I know it took the combined efforts and threats of Sir Sidney and the Colonel to convince him to hang on and fight it out behind the wall. If they had not succeeded, it would probably have been the end of the Turks in Syria for there were no other defensible places between Acre and Turkey itself, and the Turks time and time again proved themselves totally incapable of coping with the French in modern warfare. Consequently, Acre was the Allies' last chance to stop Bonaparte. As Colonel Phélippeaux later explained, "It is through this breach that Bonaparte intends to advance upon Constantinople and take the entire Ottoman Empire by storm."

Although they had convinced Djezzar to stay and fight it out behind the walls, it did not mean that he would co-operate by carrying out the agreed-upon repairs and construction. 'Stubborn' and 'stupid' are two words that always come to mind when I think of our Commandant -- the result, I suppose, of being present for much of the discussion that went on as Phélippeaux attempted to budge the old curmudgeon.

When Smith sailed into the harbor on **Tigre** -- just a few days after my unceremonious arrival -- his rank as as "plenipotentiary diplomat" helped push Djezzar into a little more action. From the moment he arrived, Sir Sidney, the senior naval officer, threw himself wholly into the land battle. His slim figure, crowned with a spectacular mop of dark, crinkly hair, was seen everywhere at all times of the day or night. His energy and enthusiasm were contagious. He had ideas and plans for everything, and must have driven the ruthless and impatient old Djezzar crazy with suggestions. Yet Djezzar showed Sir Sidney the greatest respect -- reputedly the only man he had ever so honored -- and his soldiers consequently set great store by Smith's suggestions. Sir Sidney's fearlessness is still legendary. His flamboyance could be seen in his theatrics and love of gaudy uniforms. He loved to carry out his duties clad in the most colorful Turkish uniform he could assemble. Smith approached everything with a fervor that bowled over all opposition -- whether right or wrong. Fortunately, Sir Sidney was seldom wrong. The man was fearless, flamboyant, and fervent -- a combination that can infuriate even the most patient partner.

News came from the French camp on occasion, and we soon learned they were suffering terribly from the Bubonic Plague, the same contagion that had wiped out one-third of the population of Europe during the Middle Ages. Although I had heard rumors about a "fever" while I was in Cairo, I had never realized the seriousness of the situation. Nor, do I suppose, had Bonaparte. Like some Biblical retribution, the Plague had fallen upon the French army in its fullest horror the day after the massacre at Jaffa. Stories now came to us of hundreds of French victims. All too soon, however, it invaded the fortress of Acre, and we too saw with horror the suffering of those about us. The worst part was the anxiety we each felt as we waited to be struck down by its invisible presence. The victim would suddenly be afflicted with a great fever and violent headaches. Soon a 'bubo', or boil, would appear in the groin or some other joint, growing to the size of an egg. The result was nearly always a painful death. Seldom did a patient survive, although those few who had not succumbed within four days had a good chance of making it. I alone had little fear of the terror because of my renewed faith in my Vision. Even a Fool's Paradise is better than none at all.

We learned that Bonaparte had sent a force under General Kléber northeast towards the Sea of Gallilee, where he handily defeated everything the Turks sent against him. The Druse, the Christians, the Jews, even the local Muslim tribes had long been persecuted by Djezzar, so they all welcomed the French. Rumour had it that Djezzar's enemies were collecting an army of 15,000 men to join the French in taking Acre. For the

10,000 of us holed-up behind the ancient Crusader wall, the best news was no news at all.

Of course there were some happy tidings. The first occurred on March 18 when Sir Sidney ambushed a French convoy in the harbor off Mount Carmel and captured almost all their gunboats. What made it such a monumental success was the discovery that they were carrying Bonaparte's siege-train. Sir Sidney landed these French guns along with some of our own and their crews. As a result, we soon had almost 300 guns installed in the towers and along the old wall. For once in history, the besieged out-gunned the besiegers.

"MacHugh," Colonel Phélippeaux announced one day, "you understand that you do not officially exist unless your name appears on a list somewhere. To enable you to draw rations and pay -- and not be listed as a deserter -- I have added your name to the seamen transferred to *La Sangsue*, one of our newly-captured gunboats. This is merely a paper transfer. You, of course, will remain here with me." Thereafter, to protect myself from the evil whims of our Commandant, I wore a 'uniform' of sorts. It consisted of a faded red coat scrounged from the Marines aboard *Tigre* and loose white pants -- probably naval issue. A marine's battered hat completed my ensemble. I was as scruffy as any French scare-crow, and I was uneasily aware that except for the hat my costume resembled some of theirs to an uncanny degree.

During the daylight hours the Colonel kept me busy, and even at night I ran errands or stood guard over him while he snatched an hour or two of sleep. I worried about him as the days dragged on. Colonel Phélippeaux seldom rested, and took charge personally in every crisis, large or small. The old Butcher had come to rely upon him completely. He even allowed the Colonel to build what seemed a useless interior wall around Djezzar's private garden. I watched my friend grow thin and exhausted from his intensity. He cared not a whit for his own well-being, concerning himself only with his profession and his unspoken conviction that he must exemplify the true nature of his people.

When I was able to rest, I inevitably collapsed into a deep sleep as soon as I hit the ground. Sleep did not last long, for I was terrorized by phantoms from the beach at Jaffa, or by the hissing voice of Bonaparte. The worst of all was a nightmare in which my darling Zenab appealed to me in tears. Although she was veiled I could make out her lovely features, but when I tried to go to her rescue my voice would not sound and my body would not move. Then she would place her fingers to her lips and hush me like a child. In other dreams the Sphinx gazed at me, its seemingly impassive face conveying a warning I could not understand. I always awoke shaking and in a deep sweat. Immediately fears of the Plague would swamp my mind until I cooled off. Soon I dreaded falling asleep, and like most of the defenders of Acre, I began to wander through each day in a haze of exhaustion.

Of those first sixty days only four events have ever emerged from the mists of my memory. Two involved our Commandant, the fierce old Djezzar Pasha. During the first French assault[17] the white-bearded old patriarch sat near the parapet where his men could see him, prepared to dispense rewards and punishment to his *Janissaries*. Every man to bring him the head of a Frenchman would receive a reward and recognition as a hero. Despite the small breach in our wall being over twenty feet above the ditch, General Bonaparte, with his usual contempt for his men's welfare, ordered the assault to be carried out with twelve-foot ladders. Although they had no chance, the French infantry swarmed forward. Maybe they didn't know theirs was a hopeless task, or maybe they were just full of brandy. Nevertheless, the sight of these screaming grenadiers coupled with stories of the massacre at Jaffa terrified the Turks who abandoned their posts in a stampede for the rear.

The fiery-eyed old Djezzar stepped in front of the flood of fugitives. The sight was magnificent! He was a big muscular man who dressed in plain clothes topped with a common soldier's turban. His only visible mark of office was a huge, gem-encrusted scimitar he wore at his side. Now he held two pistols aloft and cried to the fugitives, "What are you afraid of? Are you men, or are you old women? Djezzar will face them alone if you are cowards!" He jostled forward through his men until he stood alone on the parapet. With a flourish he fired his pistols down into the French who were milling about in frustration. "See! They run!" he called, and began to hurl rubble down onto them. Sheepishly his men returned to their posts in time to fire a volley or two and see the French indeed running. Thus the actions of two very different tyrants turned the tables for the entire war: from that moment on the Turks fought with confidence and determination, while French morale plummeted.

Now that many of us had reversed our opinions of "The Old Butcher", he immediately did something so heinous that even "the Young Butcher", Bonaparte, must have been staggered. During a storm which forced Captain Smith to set sail for a few days to save our ships, Djezzar Pasha assembled all the

A SECRET OF THE SPHINX

Christian prisoners he had been holding under various pretexts -- including a few French diplomats and soldiers and several hundred loyal subjects of the Ottoman Empire. There inside his fortress, despite the pleas of Colonel Phélippeaux, he had them all strangled and their corpses thrown into the Bay. In a few days hundreds of bodies were washed up on the beach near the Young Butcher's headquarters.

My third memory is more personal. The French had put enormous effort into extending a mine under the large central tower in our wall. Standing at the northeast corner facing their entrenchments, the tower outflanked their advances against either wall. The French so hated this tower that they had christened it "*la Tour Maudite*" ("The Damned Tower").[18] Sir Sidney decided to remove this mine. He landed several hundred seamen and marines and devised a plan to destroy it. Our men were to force their way into the mine itself by a surprise attack just before dawn. At the same time the Turks would assail the French trench from both sides. Once the mine had been rendered harmless we were all to withdraw. I attached myself to the marines without asking Phélippeaux, for I had begun to feel useless, walking around the fortress with him while others actually stood against the invaders.

Before dawn we assembled under the command of Colonel Douglas, who had been specially promoted by Sir Sidney so that he would out-rank the Turkish officers. We needed someone in control who was a true professional, and that man was Douglas. My own commanding officer was Lieutenant Wright of ***Tigre***. I recognized the Lieutenant as an old friend of Sir Sidney, the two having escaped from the French prison together in 1798 when Wright was a Midshipman. Armed with a Brown Bess musket and its long triangular bayonet, I joined the marines formed up in the dark. Then came a whispered order, and we slipped through a sally-port out into the real world. It felt strange to be outside the walls that had become the limits of our domain. All was silent. Only the gravelly earth under foot gave up soft shuffling sounds as we jogged forward. Far away an occasional shot could be heard, but here nothing broke the stillness.

Suddenly hundreds of voices shattered the precious silence. "Allah be praised! Allah be praised!" The Turks were launching their attack! French voices called the alarm, and a ragged fusillade of shots echoed against the ancient wall. Then all hell broke loose.

Swept along in the rear rank of marines, I found myself on the lip of a trench full of men fighting hand to hand. They all looked the same! -- until I recognized the ugly leather helmets worn by the French. I fired my Brown Bess at a French soldier below me and saw him collapse. Into the melee I jumped. A grenadier lunged at me with his bayoneted musket. I managed to dodge the thrust, and clumsily tried to retaliate, but with horror realized I did not know how to use the weapon! Bayonet-fighting requires different skills than knives and tomahawks. My opponent's unshaven face and red eyes glared at me from beyond his bayonet as he lunged again. I dropped my Brown Bess and stepped aside as his weapon slid past me. Grabbing his musket with both hands, I tried to use it as a lever to throw him to the ground. Surprised by my tactic, he stumbled and went to one knee. I heaved with all my might, and he collapsed. I threw myself upon him, but having no knife, tried to throttle him. He was strong and easily broke my grasp -- then shoved a thumb into my eye. The pain was excruciating, but the shock was worse. I recoiled with my hand over my eye, and he hurled me against the wall of the trench. Seizing his musket, the grenadier drew back to lunge at me, when a marine Corporal thrust his bayonet into my opponent's ribs. I scrambled to my feet, one eye closed and streaming. Recovering my own weapon, I searched for something I could do with more success. A party of sailors was busy tearing down the beams supporting the tunnel. I threw myself into this manual work.

Soon bosuns' whistles were shrilling the signal to withdraw. I grabbed my musket and made to scramble out of the trench when I spotted Lieutenant Wright. With my one eye I saw he had been wounded several times, for blood glistened all over him in the faint light. On the verge of collapse, he sagged against the side of the trench, unable to get out. I jumped back down and went to him. He cried out in pain then passed out as I pulled his arm over my shoulder. He had been shot twice in that arm, but I dragged him along the trench to a spot where the wall was lower. When he came to we were alone in the trench. Suddenly a pair of helping hands were extended down to us, and together we hauled him up and out. The other rescuer was a marine private. By now Wright's feet were dragging and he was unconscious again, so the Marine and I lumbered along, each with a musket in one hand and our other grasping the Lieutenant's uniform. We struggled over the uneven ground with shots whining about us when suddenly all became quiet. The French had stopped firing at us! We struggled on till we reached the sally-port and re-entered Acre.

"MacHugh! What in all ze blazes of hell are you doing here?" came a familiar voice. Colonel Phélippeaux stared at me. "I did not geeve to you the permission to join dis party! *Sacré bleu!*" His accent always grew stronger when he became excited.

"No sir!" I gasped from under my bloody load. "I just wanted to be useful, sir."

"Here, let me aid you," he said, and helped us to lower Wright to the ground. He knelt beside the wounded man and inspected his injuries. "Ah! my friend will live!" he murmured.

Suddenly exhausted, the marine and I stood gawking about -- me with my one eye streaming. An air of celebration pervaded the scene although several wounded were being attended to. Some Turks had come in with us, waving bloody *yataghans* and carrying dripping French heads by their long hair. Later I heard that sixty of these grisly trophies had been collected. We, on the other hand, had lost one officer and two men killed and only twenty-three men wounded. I believe the Turks suffered equally small losses.

I later heard how Sir Sidney had called out for Captain Oldfield and Lieutenant Wright. He had been devastated when told that both had been killed, and that Oldfield's body had been carried off by the French. He had asked for a volunteer to recover Lieutenant Wright's body, and several men had stepped forward, but the first one out the gate was a marine who coolly walked back across no-man's-land in the brightening light of sunrise. He was the one who heard me trying to get Wright out of the trench. Sir Sidney later rewarded the marine by making him one of his personal servants -- a much better 'dodge' than that normally offered a Private.[19]

Dawn was also good to me. "Well done, MacHugh," said Sir Sidney. "If ever you rejoin the 79th I shall make certain your Colonel hears of your brave deed." He kept his promise.

A SECRET OF THE SPHINX

During the last days of the siege the French attacked almost daily, suffering a fearful toll in men. It amazed me that they could keep on throwing themselves into the cauldron time and time again. Later I would realize their courage had been reinforced by brandy, served to the assaulting troops in great quantities before each attack. Nevertheless, their courage did them great credit, almost making me forget their previous atrocities. By this time they had gained a foothold in the Damned Tower, by blowing away one corner of it and occupying the bottom floor. We were now in for our most difficult time, for Bonaparte had finally received siege artillery to replace that captured by Sir Sidney. We could expect continual severe bombardment until the breach was widened. Then would come wave after wave of ferocious assaults.

On May 7, the French finally took the 'Damned Tower'. It was disheartening to see the hated Tricolor flying from the very center of our position. At dusk I stood with Sir Sidney on the wall as he scanned the enemy camp with his telescope. "By all the saints," he exclaimed (for he would never swear), "there's Bonaparte himself!" He handed me his telescope and pointed to a hill named for Richard the Lionhearted. "It's your old friend, MacHugh. Look for yourself."

There was the Young Butcher, issuing orders to several generals and pointing towards the breach. "It looks like he's organizing another attack upon the breach, sir."

Smith seized the telescope and exclaimed, "By the saints! There will be no peace tonight. We are in for a real tussle."

Just then Djezzar Pasha arrived with his entourage. When Sir Sidney related his observations the Old Butcher announced, "It is time for Djezzar to spring his trap. Let the infidels scale the breach unhindered. Djezzar will now employ the new wall built by Colonel Phélippeaux. Order the *Chiftlicks* to hold themselves ready in the second line to destroy the invaders once and for all. Djezzar shall trap them between the walls, and every last infidel shall be slaughtered!"

The deciding battle was imminent! I slipped back to our quarters. Colonel Phélippeaux had been showing signs of severe exhaustion, for his exertions had surpassed those of anyone else in Acre. Today he had collapsed on his bed. "Call me if anything happens, MacHugh," he had said as I pulled the covers over him. "I have a devil of a head-ache and am tired, but a few hours of sleep will make a new man of me." He was still asleep and would not need me till he awoke, I reasoned. His inner wall had given us this last chance to save ourselves, and I had no intention of standing by while my fate was decided by others. So I armed myself with a dagger, a *yataghan*, and a brace of pistols -- bell-mouthed St. Etienne blunderbusses, which I rashly "borrowed" from the Colonel. Still wearing my shabby 'uniform' of faded red and dirty white topped by a Marine helmet I headed for the new wall.

∽

The French were attacking! Within minutes they scaled the breach and swarmed into the gardens which lay between them and us. The eerie glow of lamps and flickering flames shredded the veils of night. Cheers and hoarse shouts of triumph reverberated as the first wave of French poured into the city they had fought so long to conquer. They knew nothing of Phellippeaux's retrenchment erected beyond their sight around the gardens. More and more Frenchmen swarmed over the breach and into Djezzar's trap.

On Colonel Phéllippeaux's wall pandemonium reigned, yet there was no panic. The blasting of canons and the constant popping of muskets made it difficult to speak. Blood-curdling shrieks of pain mixed with the cheers and shouted insults in Arabic. Above it all echoed the most haunting wail -- the eerie ululations of hundreds of Turkish women from the rooftops above and behind us. These were the women of Ahmed the Butcher's palace, and they kept up their haunting encouragement throughout the battle.[20] The *janissaries,* men of the recently-landed *Chiftlicks*, proud of their courage, and fighting in full view of these women, behaved as well as any troops I have ever seen, maintaining an accurate and steady fire upon the attackers.

"Lead us, *effendi*, and we shall follow!" came a voice behind me. I turned to see a collection of Turks, armed to the teeth. "Lead us into the gardens and we will slaughter the infidels with cold steel. We lack only an officer, *effendi*."

Armed as I was, they had mistaken me for an officer! Stunned for a moment, I looked them over. Certainly I was of no use here on the Colonel's wall without a rifle. I was armed for hand-to-hand combat, and I was here to do my bit. "Follow me then, comrades!"

We leapt down from the low wall into the open space between it and the gardens. My ill-fitting marine hat fell off and I was left bare-headed. Into the gardens we surged, my comrades hurling insults and calling upon Allah, while I screeched a Blackfoot war cry. Beside swarms of *Chiftlicks* and British seamen and Marines I led my party towards the breach.

The lights from hundreds of lanterns and fires flickered, causing strange shadows to dance and weave. Already hundreds of French, Turks, and British were fighting hand-to-hand in this strange battleground. Many small pitched battles raged for moments only to disperse into deadly cat and mouse games in the shadowy depths of Djezzar's Garden. Through it all the haunting ululations of the women sent shivers up my spine. How it must terrify the French!

We ran into a sizeable party of the enemy close to the breach. They resembled savage scare-crows -- gaunt, unshaven, eyes maddened by battle-lust and brandy. All wore faded red or brown coats and ugly black-leather helmets. Just inside the breach a huge, fair-haired general was attempting to rally his men to follow him into the garden. I fired a pistol and saw him fall.[21] We threw ourselves at his men, and both parties disintegrated into a series of individual combats. I found myself facing a wily Sergeant who had recognized me to be the *janissaries'* leader. With his long musket and bayonet, I knew he would kill me if I remained in open ground. I dodged behind a palm tree and he followed. Soon we were surrounded by a dark cluster of shrubs and palms. I took position behind a tree, giving him my chest on one side of the trunk. When he lunged I darted to the other side and threw myself at him with my *yataghan*. He could not swing the bayonet towards me because of the tree trunk. He had lunged too far. My blade swooshed down like I had practiced with Roustam and severed his unprotected neck.

I emerged into the flickering light to find three French grenadiers facing half a dozen *janissaries*, one of whom, startled by my sudden appearance, wheeled to face me. The hatred in his snarl told me he thought I was French. The *janissary* slammed his musket at my head, butt-first. I tried to dodge, but fell back against a palm. The butt dealt me a glancing blow on the temple. The last thing I saw was one of the grenadiers, an old "*grognard*",[22] firing at my attacker, as I collapsed in a blaze of pain and a burst of stars.

~

At first there was only blackness and a savage throbbing in my head. I tried to open my eyes, but the merest hint of light caused an explosion of stars and I receded again into unconsciousness. This time, however, I was dimly aware of voices and a brightness that penetrated my closed eyelids. I heard a pitiful moan, and recognized my own voice croaking, "God, it hurts. Oh God, it hurts!"

I was alive: such pain did not come from heaven. Forcing one eye open for only a second, I was rewarded with a glimpse of brightening sky. After drawing a deep breath I opened both without moving my head, dreading the agony any movement would bring. I gasped in another breath of smoky air. It was so good to be alive!

"*Sacre bleu! Il est vivant*," came a youthful voice from nearby. "What did he just say?"

"I didn't catch it, but he's lucky Uncle Pierre dragged him out. Otherwise, the Turks would have his head on a pole by now."

I froze -- the voices were speaking French!

"What regiment are you with, youngster?" came a third voice, and one of the weariest faces I have ever seen, appeared above me. Every feature was worn -- the watery pale-blue eyes, the long droopy moustache, white and tobacco-stained, the lines etched into the old face. "Come, lad, surely you can remember your regiment."

Luckily for me, all I could do was croak and close my eyes again. I asked myself 'what would my fathers do?' Both father and Sun Bull had been taken prisoner, and both had survived by using their wits. I must do the same. For the moment silence seemed the best answer.

"Don't even ask him, Pierre," said a bitter voice. "They'll just send him back to the slaughter. Let him stay here with us. He can take the day off, and we can lick our wounds together."

"I wasn't meaning to send him back into that garden," came the old man's voice. "I just thought he'd like to be amongst his comrades. He looks too much like my sister's son for me to send him back there."

At last my poor brain began to work, collecting facts and probabilities from the conversation floating around me. I was now in the French camp! My 'uniform' had caused me to be attacked by my allies and saved by my enemies.[23] Had these men heard me speak in English? Apparently not. They assumed I was one of them. I kept my eyes closed and listened.

"Madame Rumour has it that our beloved leader has ordered General Kléber to bring his regiments to join us in today's slaughter." This voice spoke with a depth of bitterness that even I found chilling. "Do you suppose Bonaparte has discovered a regiment that hasn't been decimated by the Plague or chewed up in this ridiculous siege? I didn't think it was possible."

"Don't talk like that, Marcel," came the young voice. "That sort of talk will lead to defeat."

"It is Bonaparte who has led us to defeat, boy. It is he who has thrown us again and again at a wall that has no opening. If he had waited for the siege guns we would have made short work of Acre, but no! He splashed our blood around like it was piss, till we are no longer able to do what would have been child's play a few weeks ago."

"We have all lost comrades, Marcel. Don't be so bitter!" the young voice admonished. "General Bonaparte is doing his best. He will yet lead us to victory. I hear that he announced he will personally lead the way into the breach himself and earn either Glory or Death today."

"You're a bloody fool if you believe such shit, Jean!" Marcel scoffed. "Uncle Pierre is the only old soldier left. Ask him."

"I don't bother myself about such things," growled the old *grognard*. "I know my job -- it is to fight and do what I'm told by my superiors."

"That's not why we fought the aristos and forged the Revolution," retorted Marcel.

The sounds of metal pots being scraped ended the discussion. I must have moved then because a wave of pain swept over me and I groaned. Raising my hands to my aching head, I discovered some sort of bandage. It was hard and crusty -- dried blood, mine. I felt in my belt for Colonel Phélippeaux's pistols. Only one was there! I had taken them without permission and had lost one! I was devastated. Obviously I had not yet glimpsed the larger picture.

"Yes, youngster, you are alive," a voice broke through my chagrin. "Thanks to Uncle Pierre here. He shot the Turk just as he clubbed you. Otherwise that hard head of yours would be perched upon a pole in Acre at this moment, or old Djezzar's harem would be using your mouth as a chamber pot." It was Marcel's voice.

"What's your name, comrade?" asked the young fellow, Jean.

"I don't know," I mumbled in French.

"That blow on the head has taken his senses," Jean asserted, turning to the others as though to say 'I told you so'. "What Half-Brigade are you?"

I just waved my head gingerly from side to side. It hurt too much to talk -- too much even to think. If I played dumb, I might muddle through.

"We shall call you 'Edmond' -- the same as my sister's son. You must be about the same age," said the weary old *grognard*.

"Here, try some of this soup," said the one called Marcel, putting a mug to my lips. I tried it while he went on. "We are the only survivors from that accursed garden. They say that of the first wave not one other man lives.[24] So you might as well stay here with us. I expect the 32nd Half-Brigade will be able to use one more grenadier. God knows we've lost enough. Maybe you will bring us good luck."

And so it was that I joined the *32eme Demi-brigade de Ligne*.

CHAPTER 9
"Grenadier Jardinier"
(Acre to Cairo, 9 May -- 3 June, 1799)

The fighting in Djezzar's Garden had continued for twenty-five hours without stop, and much of Bonaparte's army had been killed or wounded in front of Colonel Phélippeaux's wall. Together with those dead or dying from Bubonic Plague, the French army had suffered terribly high losses. The three grenadiers who had 'rescued' me estimated that half the besieging force had died or were disabled.[25] Disconsolate, they sat around going through the motions of soldiering. I had no choice but to remain with them.

Bonaparte had indeed brought Kléber's army to Acre, fresh from its victories near the Sea of Gallilee. That morning I heard Bonaparte harangue them, declaring he would personally lead the way and be the first man into the breach. Of course he made a show of setting off at their head, but was "restrained" by his aides who "prevented" him from leading his men into action. His performance had the desired effect, for the poor devils threw themselves into that breach like madmen, clawing their way over the rotting remains of their comrades. In front of the breach, corpses from the previous sixty days lay unburied, layer upon layer, putrefying where they had fallen. The stench was unbelievable! The cross-fire encountered by Kléber's men was overwhelming because our ships and a newly arrived Turkish flotilla under the *Capitan Pasha* joined the massed Turkish artillery in bringing down a tornado of fire on the small frontage. The new troops were slaughtered without even reaching Djezzar's deadly garden. General Bonaparte had played his trump card and lost The attack on May 9 was the last French assault on Acre.

The tale was told and retold of Captain Crozier, the young officer who had guaranteed our safety at Jaffa "on his word as a French officer". Throughout the entire siege he had put himself in harm's way, leading assaults and inviting death in his attempt to wipe out the dishonour he felt was his. That day he finally achieved his desire.

'We' had won, but I was in no position to celebrate, for I had become one of the losers. Here I was, stuck in the midst of the enemy I loathed. It was this *canaille* that had murdered my mother and massacred the defenders of Jaffa. Bonaparte's Army of the Orient was beneath contempt, and now I was a part of it! Yet amongst these monsters were three men who had risked their lives to rescue me from certain death in Djezzar's Garden, simply because they thought I was a fellow soldier. It had been 'Uncle' Pierre, the stolid old private, who had saved me and dragged me out. He had been supported by young Jean and by Marcel, the acerbic and disillusioned veteran. Since then they had treated me with kindness and had included me in their squad, although they knew nothing about me. It was only later I recognized the irony that after a two-month siege I had only one friend amongst my allies -- Colonel Phélippeaux (a Frenchman) -- while within minutes I had found three true friends (also Frenchmen) among the enemy I loathed. And true friends we were, although it took me a while to recognize the fact. For a long time I remained sullen and silent, partly out of fear, but also out of a childish resentment at Kismet which had set me down in the midst of my enemies in such terrible times.

The three *grognards* had adopted me and and now decided to name me. Uncle Pierre had already bestowed 'Edmond' upon me, but a grenadier required more. "He needs a surname," said Jean, always a stickler for legalities. I sat silently, like a stray mongrel, waiting to learn with what monstrosity their whimsy would burden me.

"'*Le Jardinier*' ('the Gardiner')," suggested Marcel. "We found him in that damned garden. Besides, it has sort of a ring to it -- '*Grenadier Edmond de le Jardinier*'."

"Just 'Jardinier'," ruled Uncle Pierre, his tired face as expressionless as ever. "It doesn't sound like we're putting on airs that way." So it happened that I was enrolled with the *32eme Demi-brigade de Ligne* as "Grenadier Edmond Jardinier".

Because they were the only survivors of their company, my new comrades were given light duties in the days that followed. Their surviving officer, being a considerate man, left them out of any of the more dangerous duties, and I was automatically included with them. The accepted story was that I had been knocked silly, by the blow to my head, and I was treated as being harmlessly deranged. That suited me fine, and I remained silent most of the time.

"You're a lucky lad to get out of there. You know what the Turks do with young fellows they capture!" commented Marcel. "Several fellows in one of the other regiments got footsore and fell behind on the march through Egypt. The next thing they knew they were set upon and gang-raped by a whole village. Some came back crying, some refused to ever talk about it again, and one poor fellow shot himself."

THE MacHUGH MEMOIRS ～ (1798 - 1801)

"Uncle Pierre"

Jean shuddered before observing, "It's horrible to think about! I can understand them doing it to women -- though *I* wouldn't do it," he added hastily. "This is a strange land with strange and terrible ways."

"Let it be a warning to you all," advised Uncle Pierre, looking around at us. "Never straggle. Stick together if you don't want to be buggered."

Re-equipped from the left-overs of the squad, I quietly discarded my British tatters in favor of French rags. Like a true grenadier, I even quit shaving my upper lip in an optimistic attempt to produce a huge moustache. In this matter I had no choice, for it was required of all French soldiers. When it had become known that only slaves were clean-shaven in the Muslim world, Bonaparte had ordered all his troops to retain a moustache at least.

My only souvenir of Acre was Colonel Phélippeaux's pistol, which I stowed in my knapsack. I vowed not to lose this one. When next the Colonel and I met, I would return it with the most humble apologies. Possibly I could capture him another to make up for my carelessness.

Morale in the French camp had hit rock bottom, and Bonaparte's name served as a curse-word. I have since read a lot of nonsense published by his admirers, stating the bald-faced lie that his men loved him. They hated him! The only thing that raised French morale during our last few days before Acre, was an attempt by the Turks to employ a little propaganda to demoralize Bonaparte's men further. Most attempts at this sort of thing that I have ever seen have produced the opposite effect, and the Turkish leaflets showered upon our trenches from the wall were no exception. "Can you doubt that, in sending you to such a remote country, the Directory had any aim other than to exile you from France and make every one of you perish?" it asked, and then went on to explain why we (I now considered myself a Frenchman -- such is the effect of shared hardships) must be expunged from the soil of the Ottoman Empire. In its final paragraph, however, this sermon offered the olive-branch -- an invitation to surrender! To my knowledge, not one French soldier took advantage of the invitation to desert. Instead, they felt insulted, and began to rally around the Young Butcher who had led them to disaster. On reading one of these ridiculous leaflets, I thought I detected the moralistic tone of Sir Sidney Smith. It seemed so like him to lecture the victims before offering them the chance to surrender. 'Get their backs up to make them refuse your offer' was what it amounted to. Smith was definitely too honest a man for propaganda work.

Bonaparte, on the other hand, was an accomplished liar whose lies have since been legitimized by gullible writers and apologists who pass themselves off as historians. To refocus his Army's hatred upon Sir Sidney Smith, Bonaparte issued an Order of the Day stating that Smith had loaded French prisoners aboard plague-infested ships and had encouraged Djezzar to massacre the Christians. These lies worked, whereas Sir Sidney's truths did not. As much as I despised General Bonaparte, I began to fear him even more. This young butcher, who had ordered his men into calamity, who had shown absolutely no regard for their welfare or their lives, had been able -- through the skillful application of lies -- to re-establish his control over them and lead them on to even greater disasters. Was he unstoppable?

After the final assault had destroyed Kléber's superb division, Bonaparte resorted to a tremendous bombardment of Acre, with special attention to Djezzar's palace, using his newly-arrived siege-guns. I shuddered when I thought what he might have accomplished if he had delayed his assaults until these guns had arrived. Meanwhile his lethargic army licked its wounds before the ancient walls and waited for the word to retreat. Finally it came. Bonaparte's 'Proclamation' was circulated on May 17. It contained a lot of nonsense meant to convince his soldiers that they had actually won the campaign. Full of bare-faced lies, it made Bonaparte and his army appear as noble defenders of Egypt, rather than the ravagers and rapists of Syria. Later I realized that this nonsense was actually meant for the French at home who wanted to read about victories and heroes -- even if they were only imaginary. Finally the Proclamation got to the point we all longed to hear -- "after capturing forty guns and 6,000 prisoners; after having razed the fortifications of Gaza, Jaffa, Haifa, and Acre, we are returning to Egypt."

Marcel's conscience prevented him from commenting on the "6,000 prisoners captured" -- and massacred -- but his sense of the ironic had not left him. "You fellows can now understand what superior beings our generals are. Bonaparte's legendary eyesight has determined that Acre has in fact been destroyed while we poor dolts still think we see the fortress standing before us. It is amazing what extraordinary faculties are granted to those who command!"

"Listen to this!" he exclaimed. "The final paragraph sounds ominous. 'Soldiers, there are more hardships and dangers facing us.' Thank God! I'd begun to grow weary of this luxury and ease. Ah-ha! I see here that our beloved Commander-in-Chief assures us, 'You will find new opportunities for **Glory**. That's good news, eh fellows! I can tell you that this soldier is in need of a bucket full of Glory. My supply has completely run out."

May 20th was memorable for us all. It was the day of our departure. However, the "destroyed Turkish army" made a major sortie from its "razed fortifications" and recaptured the Damned Tower. Savage fighting continued all day long in the forward trenches. Our squad escaped that battle. We were helping hundreds of others to carry out Bonaparte's orders to bury almost all the artillery on the beach. I took a private pleasure in this, particularly when we had to detonate piled barrels of gunpowder. That same day we received complicated instructions for our retreat. Of course, Headquarters would lead the skedaddle, preceded by captured flags and loot and by a band playing rousing patriotic marches. It would be a magnificent promenade, doubtless one of the most glorious skedaddles in all history. At eight that evening we began our "victorious withdrawal" from Acre.

By midnight we reached Haifa. There a horrifying spectacle welcomed us! In the city square we came upon hundreds of dead and dying comrades who had been abandoned there. Even before we shambled into the square the screaming and wailing sent chills up our spines. As we entered, their curses filled the air, replacing their moaning and their cries for mercy. Many had torn off their bandages and were rolling in the dust. We stopped momentarily to pick up these wrecks and carry them bodily away with us until some better method was devised.

Under these burdens we staggered to Tantura. There we discovered another 800 wounded men abandoned upon the beach. Bonaparte had ordered that these men be evacuated by the French fleet, but it had already left. Another example of Bonaparte's impossible orders -- designed to impress the citizens back in France with the hero's concern for his men.[26] "If that bastard, Bonaparte, could have swallowed his pride and contacted the British Navy, we could have gotten these poor buggers out of here." growled Marcel.

"Yes, it does seem a pity that he didn't try to do something to alleviate their suffering," admitted Jean. "They say the Englishman, Sir Smith, offered to have them transported to Egypt on his own ships. Why didn't General Bonaparte talk to him about it?"

"Because our little General is shit!" snarled Marcel. "The little bastard doesn't care about us, you fool. We are nothing! Soldiers are only bits of rubble for him to climb upon on his way to a throne. Don't forget Louis and Benedict. He had them shot as murderers of women back in Cairo. Everyone including him knew they were innocent as were all of us, but General Head-Full-of-Shit had us murder them just to curry favour with the real thieves. I'll never forget the look in Benedict's eyes when the Captain ordered us to fire. Yes, Jean, when our bones are bleaching here our beloved General will be crowning himself King Napoleon. Then he will have the whole of France to use up -- and then the whole of Europe. Our little General is the next Attila!"

Uncle Pierre said nothing, but his eyes glistened. Pierre had at last been promoted to Corporal. Eventually he spoke. "Alright, you buggers, set these fellows down," he ordered, indicating the wounded we had carried from Haifa. "Now we've got to help these other fellows somehow."

As gently as we could we set down our patients, and collapsed upon the beach ourselves. The new Corporal stood there looking over the terrible situation, trying to decide what his two lace stripes could do. "You all rest for a few minutes while I find an officer," he concluded and wearily trudged off into the gloom.

Someone must have found an officer because eventually a string of orders were shouted by the Corporals and Sergeants. To free more horses for ambulance duty, we were ordered to bury several pieces of artillery and caissons on the beach at Tantura. Sheer pandemonium erupted when one of the latter exploded, killing and mutilating a large number of those passing by. We worked feverishly to transfer the hundreds of newly wounded onto the improvised ambulances. Eventually I rose to my feet to stretch my aching muscles, and gaped with horror at the tableau. Before me was a scene from Hell! Behind us stretched hundreds of huge torches. They were the flaming remains of every home in every town through which we had passed. Around me echoed the moans, curses, and screams of the dead and dying, while a few feet away, an endless column of pale figures stumbled by, heedless of the agony around them. Only the looters and arsonists scurried about with a purpose.

That night I saw sights that defy my powers of description. Dozens -- no, scores -- of wounded and plague-stricken men were abandoned by the roadside or in the fields. They would murmur or scream, "I'm only wounded! Take me with you! I don't have the plague!" Often they would tear open their bandages to try to convince us. Some plague-victims even wounded themselves in the vain hope of being taken aboard one of the ambulances or of winning a helping hand from the rest of us. It was all to no avail. "You're a dead man," was the only comment from the weary marchers.

The bearers of a wounded officer, whose leg had been amputated above the knee, dumped their burden into the ditch while he screamed in agony. They scurried into houses to emerge with loot. Moments later these humble homes burst into flames under the night sky while the officer screamed as he bled to death only feet away. The four of us were unable to come to his rescue, as we each already carried a wounded soldier. Totally exhausted, burdened with my own equipment and an artilleryman whose arm had been blown off, I stumbled from one nameless hamlet to the next. Eventually I discovered that my gunner was dead. How long I had carried a corpse I do not know. Meanwhile the Mediterranean twinkled serenely, reflecting the flames of countless homes that mocked the agony of the Army of the Orient.

Daylight did not improve conditions. The heat was oppressive, and the horrors multiplied, for British gunboats bombarded our swarm of wretched locusts as we devoured our way slowly along the coast. It was a bitter moment when I recognized **La Sangsue** a short distance off-shore with **Tigre** standing out, firing salvos at us. On our left peasants armed with abandoned muskets picked off those they could. How they must have hated us! Smoke rose everywhere during the daylight hours, while flames and embers surrounded us at night. The fabled Plain of Sharon, one of the most verdant spots in the Levant, was being devastated on the orders of our General Bonaparte, the hero of France.

"We should never have come here," Marcel observed sadly. "This is not our world. We can win battles, but we can never conquer a land where everyone hates us so bitterly."

"It's their religion," opined Jean. "It makes them behave unreasonably. If they were reasonable they would understand how much better off they would be living under our control."

"What if they came over to France and did all this to **us?**" replied Marcel, pointing to the devastation all around, "Would you believe **them** if they said that?"

Jean could make no reply. Uncle Pierre just sat there with the saddest expression on his tired old face. I was tempted to jump in and expound on the folly of would-be conquerors and religious fundamentalists both, but fortunately refrained from saying anything.

After three and a half days of this hell we arrived in Jaffa. The bones and the ghosts of the slaughtered soldiers and their children still littered the sand above the high-tide mark. Scattered among them would be the remains of my Greek comrade of a few weeks ago, Leonidas Kapadopolis, and his tiny son. My three current comrades remained wordless as we sat there. A conscience can be a terrible burden for conquerors.

Our stay in Jaffa was mercifully short, for the next day General Boyer was ordered to march out with three-hundred lightly wounded men and the prisoners.[27] This was to be the vanguard of the Army of the Orient's triumphant return to Egypt. Once again our commanding officer pulled off some kind of a miracle on behalf of Uncle Pierre's squad. We, along with a handful of non-combatants, were assigned to guard the prisoners. What a relief to leave that caricature of an army!

Pitiful as our small force was -- almost all of whom wore bloody bandages of some description -- we stepped out with a mixture of pride and relief. We were tired, thirsty, and hungry. But at least we felt like men again. We were not part of a rabble, nor were we a pack of arsonists, thieves, and rapists. Every man hurt, but each was able to march, or at least shamble unaided. We carried with us dozens of captured Turkish flags and a pitiful gaggle of prisoners, trophies of Bonaparte's "conquest" of Syria. These were displayed in each village we passed through. I doubt if the villagers were impressed by the trophies and our tiny ceremonial guard, though undoubtedly they were impressed by the hundreds of shattered men who followed us in litters two days later, and I'm certain they remembered for the rest of their lives the swarm of rapists, murderers, looters, and arsonists who passed through their verdant land three days behind us, leaving a desert in their wake.

My emotions during this triumphal march were mixed. I was relieved to be rid of Bonaparte's Army of Death, but ashamed to be part of the display -- although only the most stupid would not see through that transparent sham. Our prisoners shuffled along, doubtless wondering what fate awaited them at the end of their long march. Of all their guards I was by far the most gentle, for I realized that if it hadn't been for a stroke of luck and an old *grognard*, I would be marching as one of them or rotting in Djezzar's Garden.

The captives were a mixed bag; they included the usual prisoners of war, plus a few sheiks taken as hostages, several Mamelukes who had loafed around Jaffa after the massacre, and a number of bewildered individuals for whom nobody seemed able to cite a crime. By far the most exalted in rank, but the basest of all, was old Abdullah Aga, Jaffa's former Commandant. I still visualized him prostrate before Bonaparte, begging that his life be spared, while the men he had surrendered were being herded down to the beach to be murdered. If Abdullah Aga hadn't been so old and weak, I might have made him the one exception in my generous treatment of the captives.

Although I found my role in this French "victory parade" to be distasteful in the extreme, there was one aspect which almost caused me to whistle a jaunty air. This was the realization that every step took me closer to my beloved Zenab. These hellish times had pushed such beautiful dreams into the background. But now, swaggering along, in relative safety, I indulged in memories and fantasies. Zenab's brown eyes often swallowed me so totally that old Pierre sometimes had to call orders twice to bring me back to the dusty track beside the sea. "I believed young Edmond was recovering his senses," I overheard him comment sadly one day, "but he seems to be losing them again as we draw closer to Egypt."

"Then he really is balmy," observed young Jean, "for he's the only one of us who isn't beginning to regain his senses as we approach Egypt. I'm looking forward to Cairo just like it was home. When I was there I hated it, but after Syria, Cairo will be heaven."

"I think young Edmond is in love," was Marcel's judgment. "Damned if he's not in love! That means he's more sane than any of us. Sacré bleu! What I would give to be in love! Do you think you'll ever fall in love again, Uncle?"

"I am a grenadier of the Republic," replied the old Corporal as though that said it all.

One day shortly after our 'triumphal march' commenced, a courier passed through. He took time to stop and water his horse, while we sat around dining on "liberated" fare and showing off our captured flags. He was bursting with gossip. "I suppose you lot haven't heard the latest orders from our beloved Commander-in-Chief," he began with assumed nonchalance. When we acknowledged our state of ignorance he plunged into his story. "It was the other day in the hospital -- you know, the one in Jaffa. Well, General Bonaparte visited the wards. He tried to rally the men by telling them the Turks would soon be arriving. There were fifty or so still in their beds -- mostly fellows down with the plague. When they didn't show any interest in getting up and onto the road, he ordered the doctors to poison them. Can you believe it, comrades?" Murmurs of disbelief greeted his claim. "It's true. They were each given enough poison the finish them off, and we left them there for the Turks to bury."

"That's ridiculous!" scoffed young Jean. "The General would never do something like that to us. We are his soldiers."

"You can believe whatever you like, youngster, but I was at the door waiting for this dispatch when I saw it all," replied the indignant courier. "I'm not the only one who saw it either. There were officers and doctors there. Well, to hell with you lot! I'm off. My advice is 'don't catch the plague or he'll poison you too'."

"That's pretty hard to believe," said Jean after the courier left. "General Bonaparte wouldn't order us to be poisoned if we get the plague, would he?"

"The General is above nothing!" growled Marcel. "I suppose he thinks it would look bad in the newspapers back home if any of us were to be captured -- or worse yet, **saved** -- by the Brits or the Turks."

"Maybe he was trying to save them from the pain," suggested Uncle Pierre, "-- and the lonely death."

Once the courier's rumor spread there was a good deal of such discussion, but the prevalent emotion was anger directed at Bonaparte. These men who had carried out his orders to slaughter almost 6,000 defenseless soldiers and children -- who had thrown themselves into completely hopeless assaults against Acre for him -- who had been the victims of his neglect throughout an entire campaign -- were now enraged because they believed the man had ordered the most humane way to end the suffering of fifty hopelessly ill comrades hours away from being captured and probably slaughtered by the Turks![28] It passed my understanding. I was very young, so I believed Bonaparte had already shot his bolt -- that he could never again hoodwink an Army -- let alone a nation -- into following him in his search for that elusive ephemera, **GLORY**.

We marched for two whole days across the Sinai Desert. It was exhausting, yet exhilarating -- broiling hot during the days, and shivering cold at night. But oh! Those nights under the stars! The clean desert air must act like an enormous magnifying glass, because never have I seen so many stars, and never have they sparkled so brightly. And the silence! It made one whisper, reluctant to mar the perfection of it. Nights on the North American prairie had captivated me, but there had always been background sounds -- coyotes, owls, the rustle of the long prairie grass. Here there was no sound at all, no life. While the others slept, I lay gazing at the heavens, but seeing my lovely Zenab, imagining her voice. When at last I slept, her face remained. Gone were the nightmares of Syria.

One day a stunning phenomenon occurred. Marching along in perfect weather and bright sunshine, we became aware of a strange rushing sound, and for a moment the sun appeared blurred. Then all at once it vanished! Over the ridge before us, leapt a great sandy curtain. Within seconds we were lashed mercilessly by an incredibly savage wind wielding millions of barbs of sand, each one a cutting edge. Our column of heroes disintegrated instantly as everyone sought cover. An hour later we resumed the march in bright sunshine, marvelling at the suddenness and violence of Nature's ambush.

Our march at last took us into Egypt, then on to El Arish, where four months ago these same men had murdered every inhabitant. There was not much left in El Arish now, but Egypt will never be short of people, so already the ruins of centuries were coming back to life with new inhabitants. Once across the border, a great weight dropped from every soul. Even Marcel showed signs of joy as we neared Cairo. Forgetting they were foreign conquerors in a hostile land, the soldiers behaved as though they were returning home. Our poor captives appeared no happier, however, for none knew what lay ahead. Only that evil old coward, Abdullah Aga, could count on survival. He had Bonaparte's word for it. He had already paid for his life with his honor and the lives of his men.

Partly through fear, and partly through bitterness, I had remained aloof from my new comrades through the weeks I had been with them. Although I had by now developed a grudging affection for all three, I carried on my performance as a taciturn lost soul, a man without a past. I hated all they stood for and abhorred the atrocities they had committed, yet while I had marched with them, their humanity and courage were all that I had witnessed. Old Pierre, bitter Marcel, and the trusting Jean -- they were my best friends. Nevertheless, I determined to slip away from this life I hated. My goal was the daughter of Sheik El Bekri and abandoned mistress of the Young Butcher -- the lovely Zenab.

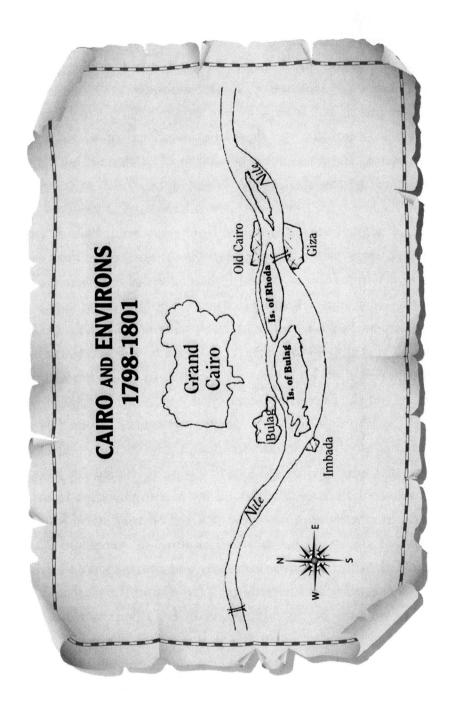

CHAPTER 10
"Caspar's Return"
(Cairo, 14 June - 15 July, 1799)

On June 14 the Army of the Orient made its triumphant entry into Cairo. I marched with it through the Bab el-Nasr, "The Gate Of Victory". There were not many of us, for Bonaparte had included only the healthy survivors. The hundreds of wounded and sick had been squirreled away in various villages and towns so the locals could not calculate our losses. The four of us marched as one rank, the remnant of a company of *le 32eme Demi-brigade de Ligne*. Like everyone else in the charade I wore a palm frond in my ugly leather helmet. We had remained outside Cairo for several days until the main body caught up to us. During that time we had eaten like kings, washed our filthy clothes, and in general tried to forget what we had seen and done. Now looking more like soldiers than scarecrows, we strutted through the Gate of Victory into a city which has seen so much pageantry over the centuries. Palm fronds had been strewn about the streets in front of us, bands played jaunty airs, and a huge crowd turned out to assess us. The lucky garrison that had been left behind to guard Cairo presented arms to welcome us, the notables marched with us to Esbekiya Square, and the curious populace crowded the streets to watch. Marcel, of course, wondered aloud how our "Triumphal Entry" compared to the hundreds staged in those same streets through the ages.

We "Victors of Syria" were then given a few days off to rest from our exertions. Our uniforms were in tatters, despite our efforts to gussy-up for the big parade. Morale was almost as bad. The true state of affairs can best be understood by reading Bonaparte's orders concerning "agitators". Every battalion was ordered to make a list of such negative-thinkers and forward it to the Commander-in-Chief. Every time an "agitator" was convicted of some infraction his punishment was to be double that normally handed out. In some circumstances the 'agitators' were to be shot without trial. These orders did not make the army happier -- only quieter. I know that Marcel, gritted his teeth and shut his mouth, but it was probably too late to keep his name off the list.

This period of lessened duties and relaxed discipline was the ideal opportunity for me to "desert", if that is what it can be called. By now "Grenadier Edmond Jardinier" had unwound to the point where he appeared to have become a regular member of the squad. I had warmed to these three personable characters. Leaving them was going to be a sad experience, although the hope of seeing Zenab eased the pain considerably. Consequently, one morning I set off by myself, allegedly to see the city and maybe even the Pyramids. I hoped that way, Uncle Pierre would delay reporting me missing when I did not turn up that night. Young Jean wanted to come along, but I declared my intention of going alone. In my knapsack carelessly slung over my shoulder I carried my few possessions, including Colonel Phélippeaux's remaining pistol.

Although my initial impressions of Cairo had been negative, I had developed an affection for this exotic, squalid city. Despite their villainous appearance, I had even grown to admire the cheerful fatalism of the *fellaheen* and the lower orders. Tucked away in corners of this sordid city lay islands of beauty and tranquility, and my recent experiences had made me relish these finer attributes of Cairo. Therefore it was with a light step that I set out for Sheik El-Bekri's palace where I waited outside the gate used by the Mamelukes. For once, fate befriended me. I had only been there a few minutes when out rode Roustam Raza!

"Roustam, old friend," I called.

He looked about him but saw only a ragged French grenadier with a straggly moustache. When he showed no sign of recognition I advanced and asked quietly, "Do you still stand guard over the General's brown-haired lady?"

Roustam stared down at me from his mount. "Caspar?"

"Yes, it is I."

He looked bewildered, but cannily said nothing.

"I must speak with you ," I insisted.

Still without any word of greeting, he looked about before surreptitiously pointing towards the small park a short distance away and rode off. A few minutes later I joined him there and told him what I wanted. "I wish to resume my life as Caspar the Mameluke. Can you get me back into the house, Roustam, old friend?"

"Yes, I can let you in the gate. After that you are on your own. The master has recently found some new confidant and seldom calls me to his chamber,"-- this with a bitter edge to his voice -- "so I no longer have the influence I had when you left.

I dare not be discovered helping you. However, your room has been left as it was. Your absence has gone un-noted -- except by one. I explained to her that you had been sent on a mission of great delicacy. She pouted, but had no choice but to accept." My face must have revealed my thoughts for he laughed. "My poor Caspar! She has a new guard outside her door these nights -- a devilishly handsome fellow with flashing eyes and a magnificent moustache -- and a great swordsman, so I am told."

I was stunned, for I thought I knew the identity of my replacement. Roustam pretended not to notice. He looked me over distastefully and added, "I shall go to my room and bring something to cover those rags. Then you badly need a wash and a change of clothes -- and Caspar, do shave off that ridiculous moustache. Aboonapart's grenadiers are supposed to grow handsome manly moustaches." With a loud laugh my old friend remounted and rode back towards Sheik El-Bekri's palace.

Within half an hour I was slinking through the corridors of the mansion on my way to my own room. Grenadier Edmond Jardinier had deserted. Caspar the Mameluke had returned.

~

That evening I put a bold but foolish plan into operation. It was dazzling in its simplicity -- I would turn up for duty as though I had never been away. Through the familiar corridors I swaggered -- early enough, I thought, to arrive before my replacement. I was mistaken. When I turned the last corner I saw him at the end of the hall, posturing and swaggering arrogantly before Zenab's door -- Omar of Myra!

Omar of Myra was athletic, but I was sure I could handle him physically. I remembered, however, Roustam's warning about his swordsmanship.

"You can go now," I announced abruptly to him as I arrived. "I have returned to resume my duties. Thank you, Omar of Myra, for your diligence in my absence."

He appeared dumbfounded, but reacted as I had feared. "Begone, dog! Omar of Myra now stands guard over Sheik El-Bekri's daughter. Be off or you die!" He stepped forward threateningly with his hand on the hilt of his scimitar. I couldn't risk letting him draw that blade, so I surprised him with a hard right to the belly. He doubled over in pain, and I caught the look of astonishment upon his mobile features. As his head went down, it met my left in an uppercut which I threw with all my might. The crack startled me, and pain shot up my arm. His head flew back so hard that his cahouk came loose and fell to the floor beside him. He lay huddled and unconscious before me.

I stood rubbing my aching knuckles, wondering what to do with him. I had never knocked anyone unconscious before. If I left him there to come around, he might make a scene, but I would at least know what he was up to. If I could haul him away somewhere, he would be spared the humiliation of being discovered or of facing me -- and Zenab. That might make him less inclined to challenge me later. I chose the latter course, and carted him off to his room. There I hoisted Omar off my shoulder and dumped him onto the one great bed in the chamber. From the midst of the satins popped a face. The eyes were enormous and the features were like those of a large baby -- pretty but without expression. I shuddered involuntarily when I realized that it was the face of a man.

Outside Zenab's chamber I stood on guard once again -- just like the old days. After what was probably only minutes I could wait no longer. I opened the door a crack and slipped into the darkened room. The scent of flowers and the rustle of silk released a flood of wonderful memories. Then I heard Zenab's gasp of surprise.

"Caspar!" she whispered in amazement. "Is it truly you?"

"It is," I assured her. My eyes were growing accustomed to the darkness as I slipped my weapons and clothes to the floor.

"Roustam said that you had gone far away, that you would never return. I wept for you, Caspar. I longed for your love."

I slipped under the covers beside her and without a word let her know how deeply I had missed her.

~

The next day around noon Roustam came to visit me in my room. "I came by last night to see you and discovered you were not here in your chamber," he said with a malicious grin. "You couldn't have been with my master's daughter, for she was protected by none other than Omar of Myra, the great swordsman." He waited expectantly, and I could see he was not going to leave so I told him briefly what had happened.

"You must watch your step, Caspar. Omar of Myra is not one to trifle with. He has killed before -- but never by acting rashly. He will bide his time and wait for an advantage; then he will strike -- and his method is unique," he warned. "Now come, let us go riding."

I enjoyed being astride a horse again after my months as a footslogger. We ventured out without the battalion of stick-wielding ruffians usually hired by men of consequence. I suggested a visit to the Sphinx. I had no other reason except idle curiosity, but that enigmatic face drew me like a magnet, and once again I had an overwhelming sense of foreboding when I touched it. Standing upon our saddles, Roustam and I clambered onto the creature and strolled along its back.

Eventually we climbed the great head and sat looking down upon its brow and the shattered nose. A vague tension came over me, much like I used to feel upon encountering a burial site on the prairies.

Roustam was relaxed and seemed to be enjoying himself. Yesterday, he had been irritable and sullen -- a result of his loss of favor with the master. On the other hand, I was basking in the joy of my night of love. "Tomorrow I am to join my master, the Nahib al-Ashraf, for an audience with the great Aboonapart himself," he said with satisfaction. "That is most certainly a sign of my master's pleasure. His new companion is not to be present, so he must have fallen out of favor. I admit I am not surprised."

This gossip interested me only slightly: There were other matters about which I was burning to enquire. "Is Aboonapart now in residence at our master's house?"

"No. Since he returned from his conquest of Syria he has resided at the palace of Mohammed Bey el-Elfi." He paused and looked innocently towards the Pyramid of Cheops. "Were you with Aboonaparte on his glorious campaign of triumph?" he asked casually.

This was not a subject I wanted to reveal to Roustam Raza, so I brushed lightly over it. "No. I served the Sublime Porte, but I was captured. Because I speak French, I was mistaken for a Frenchman, and was forced to march back with them. There was no 'triumph' nor 'conquest'. Aboonapart was totally defeated. His days here are numbered." Now it was my turn to put a question I had been dying to ask. "How did you come to be working for my boss, Mister Flemming?"

This time it was Roustam who became evasive. He sat looking out over the desert deciding how to answer. "I did not know the man's name. Flemming, you say?" When I nodded, he repeated the name several times as though memorizing it.

I knew then that I had made a serious mistake. Roustam was not lying. He had not known Flemming's name. Had he tricked me into revealing this vital clue, or had I just dropped it like a fool? The last time I had been here with Roustam I had been so surprised by our encounter that I failed to watch him for 'give-aways'. Had he lied to me then? I would never know! From here on I had to very careful and watch both of us for 'give-aways'.

"Well, how **did** you come to be doing that sort of work?" I persisted. "And why did it take you two days to deliver the message to me?"

"I do not do 'that sort of work'. When I met you here I was only doing a favor for another who had asked me to meet you because he was unwell." Roustam stared off to the horizon for a second before stating casually, "I do not know how long he kept the letter. I brought it only as a favor to him. I will not be doing it again."

A SECRET OF THE SPHINX

He was lying. There was no point in continuing the questioning. I had given away a lot of information in exchange for nothing. Roustam was a better courier or agent than I. When we rode away it was my companion who was cheery, while I felt downcast and foolish. I even forgot to look back at my friend, the Sphinx. If I had, I'm sure I would have seen a warning of what was to come.

A day passed before I saw Roustam again. Shortly after dawn when I returned to my room after a night with Zenab he was waiting for me. My friend was more than crest-fallen -- he was shattered! His jovial face was ashen, and traces of tears streaked his cheeks. Every line of his body displayed his dejection. He looked up at me and sobbed, "My master has given me to Aboonapart."

~

Over the next few weeks several incidents occurred which I can no longer put in their correct order. Living in the never-never land of an Egyptian palace, balancing the roles of Mameluke bodyguard, French lover, and British spy, all the while watching over my shoulder for a murder attempt and trying to keep out of the way of both General Bonaparte and the Sheik El-Bekri must have disrupted my sense of time and order.

I kept my ears open and discovered that Roustam had indeed been given as a present to Bonaparte along with a great black stallion. It was easy to understand my friend's anguish, for Bonaparte had wanted Roustam killed after he had witnessed the embarrassing scene in Madame Fourès' chamber. On the other hand, as Bonaparte's private valet, he would no longer be a threat to the General. But what if my friend really was in the pay of our Secret Service? He was now at the very heart of the enemy camp. What a great source he could be! Unfortunately, we no longer resided in the same mansion, but I determined not to lose touch with Roustam.

Soon after our return to Cairo Bonaparte had ordered the execution of all the prisoners we had escorted from Syria. They received no trials, and the excuses cited were trivial. I know that several Mamelukes were shot for having failed to purchase safe-conduct passes from French officials. Eventually, to save ammunition and reduce the gunfire which might alarm the citizens, the Young Butcher ordered the majority to be beheaded. One of the last to be beheaded was my cowardly old commandant at Jaffa, Abdullah Aga. Inwardly I smiled when I recalled his tears of joy at being promised his life by the 'Young Butcher'.

Zenab and I continued to make love each night while I pretended to guard her virtue. As the weeks passed, we slipped again into a pleasant routine. I recognized that my earlier passion was waning. Physically, Zenab was as alluring as ever, but we shared few interests, and our outlooks could hardly have been more disparate. To an Egyptian or Turk these would not have mattered, but I wanted a companion as well as a lover. For her part, Zenab still yearned for Bonaparte. Although she never actually said so, many little things and casual comments suggested that she was waiting for him to "regain his senses". In the meantime, we enjoyed the love-making -- although there were times when I worried that I was there only as a substitute for the Young Butcher.

Gossip was the main occupation in Sheik El-Bekri's palace. The hero of the day -- although no one would acknowledge it within hearing of the French -- was the rebel, Murad Bey. I had to admit that I longed to join his rebel band. Tales of his daring were the talk of the bazaars, and he sounded like the sort of man I could follow. My love-life no longer presented any apparent danger, and there seemed no possibility that I would ever be called back into service as a courier for Mister Flemming or Sir Sidney Smith. In fact, they could not know I was still alive -- let alone that I was almost a brother to Bonaparte's new valet. Meanwhile, Murad Bey swash-buckled about Egypt carrying out lightning raids on the French without ever being caught. Everyone knew of his successes because nothing remained secret in Cairo -- a fact that should have frightened me. Consequently, one morning the city buzzed with the news that Murad Bey had 'visited' Cairo last night. He had outwitted the troops sent to intercept his band, and had scaled Cheops' Pyramid, from the summit of which he had flashed signals to his wife which she had returned from the roof of his mansion in Cairo.[29]

I had been trying for weeks to convince Zenab to venture out of the city to visit the Pyramids and the Sphinx. It would take much preparation, because the usual gang of stick-wielding ruffians had to be hired to clear the way through the streets, a camel rented and equipped with an enclosed basket for Zenab to insure she was not visible, and arrangements made for servants and guards. I was willing to make the arrangements because it would be worth it for Zenab. She had always drawn back from committing herself to the plan, so imagine my surprise the morning after Murad Bey's escapade, when I was summoned before her shortly after I had returned to my room.

"Caspar, I wish to see the spot that everyone is talking about. I wish to visit the Pyramid of Cheops today." she announced.

Now it was my turn to be reluctant. "But I cannot make all the preparations on such short notice," I protested.

"Nonsense! The preparations will be simple. I will ride with you -- disguised as a man!"

I was flabbergasted. Zenab had shown occasional bursts of energy and daring, but this was beyond anything I had expected.

"I have decided," she declared. "You shall get me a Mameluke's uniform. One of yours will do just right. I would like the green yalek you wore two nights ago -- and the red charoual. Get me a powerful charger and we shall gallop through the city like two warrior princes. Oh, it will be so wonderful to be a man and throw off all the chains of my life!"

"Have you ever ridden a horse?" I asked, to bring her back to reality.

"No, but it cannot be very difficult. I have watched from the roof and have seen the Mamelukes and the French ride their horses, and it appears to be as easy as sitting on a cushion."

And so we rode out of her father's palace shortly after noon. Our style was less grand than Zenab had envisioned, but was splendid, nevertheless. She sat precariously upon a gentle little bay mare. It was not uncommon the see Mamelukes twelve years of age and even younger, so her size and face would not draw comment, I hoped. We had padded her tiny waist and made certain that her breasts and long hair would not give her away. My red and green outfit was a bit baggy for her, but padded in the right places, it suited her just fine.

Today for some reason the courtyard was empty of the French officers who usually lounged about. Because I had decided to take the least convoluted route, rather than spend more time jostling through the city itself, we picked our way through the crowded streets of Cairo, then rode the mile to the small city of Bulac on the banks of the Nile. Surprisingly, there were few French about anywhere. Possibly, I reflected, they were out chasing the elusive Murad Bey. Zenab tended to be a bit imperious, so the boat journey across the Nile was nerve-wracking for me, as she took it into her head to hector a lazy boatman.

Because we were forced to ride at a walking pace, it was late afternoon when we topped the last rise before the Pyramids, their peculiar golden hue heightened by the setting sun. I was stunned by an amazing sight. The entire area swarmed with French troops! Battalions of infantry drilled before rows of tents while couriers galloped back and forth. The 'Sultan Kebir' had decided to take on Murad Bey himself I reasoned. As if on cue, trumpets blared from before a large tent topped with a French Tricolor, and out stepped General Bonaparte.

We stopped on the crest of the dune to gape at all this activity. I stared hard at the Sphinx, but it did not respond to my unspoken query. Suddenly Zenab dug her heels into the mare and it shot off down the slope. It was a miracle she didn't fall off, for she swung wildly from side to side before gripping the mane and hugging the great bay neck. Several hoots of derision from French cavalrymen standing beside their horses, warned me Zenab had been spotted. I urged my grey down the slope with a conscious effort at grace. As I came to a halt beside her, Zenab slid awkwardly from her mount and stepped off towards the soldiers, abandoning her mare. I swung down, drove my scimitar into the sand, and hurriedly tied the two sets of reins to its hilt. Hopefully the horses would still be here when I brought her back.

My little Mameluke was discovering how difficult it can be for a novice rider to walk, so it was easy to catch up to her. Trying not to attract attention I demanded in an angry whisper, "Where do you think you are going? Turn around and come back to the horses."

This was the wrong tack to take with Zenab. "*I* decide where we go," she hissed. "You are just a servant -- **my** servant. I wish to see the soldiers up close."

These troops still wore the patched rags from their Syrian campaign, but there was a new spirit of optimism, and even in the grasping sand, they swaggered and strutted. The Young Butcher had already erased from their minds the failure at Acre and the terrible retreat. Once again they were prepared to follow him to 'Glory'. I wondered if my three friends of the 32nd Half-Brigade were here. Not that I feared recognition by them. No one would relate this flamboyant Mameluke with the sullen Grenadier Edmond Jardinier. "Slow down!" I ordered my companion quietly. She saw the sense in that, or possibly her legs were tiring in the deep sand, but she slowed the pace so we could saunter along like curious professionals looking over our fellow soldiers. When I recognized where she was heading it was too late.

Suddenly we emerged from the crowd before the large tent flying the Tricolor. Bonaparte stood several yards away looking through a sheaf of papers with his staff officers. Lengthening her stride, Zenab walked straight towards him! A great warm smile covered her face. To me it was as if her disguise had been torn away. I looked around us, but incredibly, no one seemed to notice her approach.

A SECRET OF THE SPHINX

As I reached her, Zenab suddenly froze in her tracks. I seized her arm, but my eyes followed her glare. Emerging from the giant canopy was a vision of loveliness, Madame Fourès, the beautiful "Bellilotte", clad in a hussar's jacket and tight blue breeches. Zenab shook off my arm abruptly, and her hand emerged from the folds of her green robe clutching a small dagger. I seized the arm again and swung her around with as little fuss as possible. Her eyes now blazed at the officers. It was Bonaparte she wanted to kill!

Trying to appear like two friends chatting, I forced her hand downward Now she glared up at me, forcing me to exert my strength to subdue her. The small dagger fell to the ground between us, but I didn't let go of her wrist. Looking down casually, I shuffled my feet in the soft sand to cover the blade. When I looked up Bonaparte was gazing at us, a puzzled expression on his face. Zenab glared insolently back at him. "Welcome to our camp, comrades," he said graciously. "I see you have come to help us track down the bandit, Murad Bey."

Zenab continued to glare angrily at him, so I tried an answer to divert his attention from her. "I regret, my lord, that we must remain with our own master, but I'm certain our absence will not deprive you, as you have many men here to capture one bandit."

Bonaparte ignored my comment, and stared at us. "I have met you before -- both of you."

"We are of the house of the great Sheik El-Bekri. We are honored that the exalted Sultan Kebir has remembered simple servants such as ourselves." I replied, bowing and forcing Zenab to make a jerky, graceless bow.

The General continued to stare, focusing now on Zenab. "No," he replied slowly, "I have met you both somewhere else. Never mind. I shall think of it presently." With this he turned to his staff officers and began to study another set of papers.

Holding Zenab's arm tightly, I turned her around and began to walk back towards our horses. "Thank you, God," I whispered in English.

"Just a minute you two!" screeched a voice. It was Bonaparte. "Turn around! Let me have another look at you. Ah! It is as I said -- I **did** meet you both earlier! I never forget a face. Berthier! Do you not recognize these two? Come, come! You have seen them in circumstances dramatic enough. Bessiers, surely you recognize these two!" he cajoled his officers, enjoying everyone's embarrassment, and his own control of the situation. He sneered at us triumphantly. "You are two of the Mamelukes presented to me that night after my victory at Jaffa. I sent one of your comrades with my proclamation to Djezzar Pasha, the Butcher of Acre."

~

That night there was no love-making in Zenab's chamber. The poor girl was suffering from saddle-sores and aching muscles. The pain did not dim her excitement, however. It was clear to me that our little excursion had been the highlight of her life. She babbled on constantly about the thrill of venturing out in men's clothes, living as a man, seeing the greatest wonders of the world -- of fooling Aboonapart. I was both relieved and saddened by a note I detected in her chatter -- a hint that this would never be repeated, that she would be satisfied to live with the memory of that day.

"Were you actually going to kill Aboonapart?" I asked.

"Of course! The sight of that bitch parading as though she were a soldier made me so angry."

"But Zenab, you were there parading as a soldier yourself."

"But she was there with his blessing! I had to sneak out of my father's house. Aboonapart would never let me dress like that and appear in public with him. It is Aboonapart whom I wanted to kill."

"Yet you also want him to return to you," said I in a moment of clarity. "You love him more than you hate him, don't you?"

After a moment of silence she responded softly. "Yes. You are right. I could not really kill him." Another pause before she stated resolutely, "I shall take him back from that French whore. He and I shall rule Egypt and have many children, and our children's children will rule Egypt for a thousand years."

This sort of talk did nothing for my heart nor my self-esteem, so I spent the next hour or so in a state of humiliation and shock, while Zenab chattered on cheerfully about the wonderful sights I had shown her today and her hopes for the future. It was her night, and I sat on the edge of her bed mumbling replies, marveling at her beauty, and silently lamenting the loss of my dreams. My hatred for Bonaparte knew no bounds. Although I could not let Zenab do it, I would gladly kill the Young Butcher if ever given the chance again.

It was relatively early when I returned to stand guard outside her door. I had only been there for a moment or so before I heard stealthy footsteps coming from one of the side passages. I stepped back into the shadow cast by a gigantic plant in an urn beside the door, and peered through the fronds as the steps drew closer. A *cahouk*-covered head peeped around the corner, but it was too dark to distinguish the features. Presently two bodies emerged and tiptoed to the door. It was Omar of Myra and his *serradj*, Eznik, the little Armenian with the baby face. Eznik was gingerly carrying a sack which he held well away from his body. Omar bent an ear to the door, but heard nothing.

I stepped forward. "Do you dare to intrude upon the lady's privacy?" I demanded softly. Omar spun around, caught completely off guard. He held no weapon. Thus my impulse to strike him dead was stayed. He straightened and looked me in the eye. "I know your secret, Caspar. You will pay with your life for striking me." His voice was soft, but it rasped with hatred. "Your time will come soon, dog!" Something thrashed about inside the sack held by Eznik. Omar glared at me for a moment then wheeled and departed as silently as he had arrived, followed by his *serradj* holding the sack at arm's length. Omar had probably been at the door every night to eavesdrop on our love-making. Which secret did he know? I had so many secrets to hide!

I puzzled it out that night -- almost a welcome relief from the anguish I felt over my fatally wounded infatuation. If Omar wanted to have me executed for sharing Zenab's bed, he would have done it weeks ago. But he could hardly report me to Zenab's father without revealing his own failure to guard her. Besides, I suspected he wanted her for himself -- despite his Mameluke preference for his *serradj*. If Omar of Myra revealed my transgressions he would also destroy Zenab. No, there must be something else he had discovered to hold over me.

I was still standing at my post, immersed in gloom, when I heard another set of footsteps. These were shuffling rather than stealthy. I soon made out the form of a servant. It was the eunuch who had approached me months before with the instructions that had sent me out to the Sphinx. My heart skipped a beat.

"It is dawn. Your duty here is done," he said quietly. "My mistress orders that you come immediately. There is more important work to do. Follow me now, please -- *moallem*!"

I followed him in a daze. The term '*moallem*' was reserved for an educated Christian, and although it had been spoken in irony it meant the eunuch knew I was not a real Mameluke. Eventually we arrived by a circuitous route at the harem of Sheik El-Bekri. All was quiet, for the ladies of the house were not early risers. We slipped into an alcove and the eunuch disappeared, leaving me in the dark. But I was not alone. A soft hand took mine and pulled me gently down onto a bed of silks. Naked arms and legs twined about me, and two ample breasts pressed against my face. A voice whispered, "You have been neglecting me. I have need of you my young lion. But first we make love."

Jasmine was very persuasive, so the sun was shining brightly before she spoke again. Satiated at last, her languid brown eyes gradually returned to their normal laughing sparkle. "Ah," she sighed in satisfaction, "my daughter has chosen well. It is a pity you are an infidel, but her secret is safe with me," she whispered.

I was stunned. I tried to think of something to say -- a denial, a protest of some sort.

"Do not fret, little blue-eyes," and she kissed me lightly as she continued, "Your secrets are safe. It is my little girl about whom I worry. Zenab is so gentle and too vulnerable to have been placed in the situations she faces." Her words were spoken in a soft murmur. "Oh Caspar, I wish it was not so late. It would be so wonderful for us to do it again!" Jasmine sighed and stretched her plump loveliness lasciviously. Suddenly she raised her head as though listening attentively then stared me in the eye and whispered, "The nightingale's song is so sweet, isn't it?"

My first reaction was 'There are no nightingales here'. Then the true meaning of the words struck me, and I mentally fumbled about before I could remember my reply. "Yes. But only in summer," I stuttered.

She smiled enigmatically at me before continuing in a soft whisper. "Now, my young Englishman. You must leave me and take this message to Mister Flemming. Yesterday a Turkish Army landed at Aboukir. Regrettably, Aboonapart has already heard the news and prepares to march there to defeat them. If he moves quickly, he will crush the Turks before they can prepare for battle. Yesterday, unfortunately, he assembled his soldiers to hunt for Murad Bey, so he is already one day ahead of what was expected. Mister Flemming must be told. Every hour gives Aboonapart an advantage."

All this was so softly spoken and so interspersed with kisses that anyone listening would have mistaken it for the preliminaries to love-making. Thrusting my face into her breasts again, Jasmine continued. "I have no one to send with you to guide you this time. My couriers have all disappeared -- all seven. You will have to make your own way to Aboukir. Wear your special sash and find the Turkish soldiers who have landed there. They will take you to Mister Flemming."

There were too many questions surging through my mind, so I kissed her and whispered into her ear, "Zenab, does she --?"

I didn't complete my question, for Jasmine pulled my head around by the ears and whispered, "She knows nothing! Do not even think her name! She must be kept out of these affairs. Now go!" She thrust me away as though disappointed with our love-making, and rolled over in a silent mock tantrum, Hastily I arose, acting out my role as the rejected lover. In complete silence, I strode out with the feeling that several pairs of eyes watched my performance.

CHAPTER 11
"More Psychological Warfare"
(Aboukir, 20 July - 5 August, 1799)

"Frankly, MacHugh, I never expected to see you again," said Sir Sidney Smith, a smile warming his bird-like features. I had just been brought aboard **HMS Tigre** anchored in Aboukir Bay. "You were listed as a deserter several days after your mysterious disappearance. However, your reappearance here would seem to cast some doubt on that theory. You do know the punishment for desertion, don't you, MacHugh?"

I gulped. "Yes, sir. Death." I was dumbfounded. "I didn't desert, sir. I was captured." I launched into an account of my part in the battle at Djezzar's Gardens, my 'capture', and my return to El-Bekri's palace.

"Well, MacHugh, I'll discuss your story with Captain Creevey, commanding officer of **La Sangsue,** before he decides upon any disciplinary action. You must realize, MacHugh, that you cannot simply run off and take action on your own. You are officially a common seaman, forbidden to make such decisions. It was commendable of you to throw yourself into the fighting in the gardens at Acre, but it was totally without authorization. Your commanding officer will tell you when to fight. Until that time you must control your ardor. Is that understood, MacHugh?"

"Yes, sir."

"Having officially dealt with this case of alleged desertion, I must now commend you on your initiative in surviving the ordeals you have endured. Have you any questions?"

"Yes, sir. Several."

"Go on, MacHugh. Ask away. As you probably realize, I have formed an attachment to you because of your daring and your independence. I have always leaned that way myself. Nevertheless, I can advise you that life proceeds in a much more orderly and placid manner if you will but follow the rules laid down by society, and accept the judgment of those placed above you." This was strange advice from the ultimate man of initiative I thought. "I will do my best for you with Captain Creevey. Now ask away."

"Sir, how is it that this Captain Creevey is my commanding officer? I have never heard of him. I am not a seaman; I enlisted in the 79th Highlanders. Since then I have worked for Mister Flemming and for you and for Colonel Phélippeaux."

"At Acre you were assigned to the gunboat **La Sangsue** which is commanded by Captain Creevey. You have to draw rations and pay somewhere, you understand." I nodded although I had never drawn any rations and had yet to be paid a penny for my months of service. "Mister Flemming is a civilian, thus unable to command anyone officially. I am in overall command, I suppose, of the nefarious schemes in which you have become embroiled, but I am not your commanding officer as such. Poor Phélippeaux had no official British status at Acre. His commission was with the Turkish Army." He noted my puzzled expression at his use of the past tense and added, "Did you not hear about the Count? No, of course you couldn't. I'm sorry to tell you he died the day after you disappeared. You and I were dear friends of Colonel Phélippeaux, so I feel I can say this in confidence to you, MacHugh. Officially I have attributed his death to overwork and exhaustion, but in truth he died of the plague -- a most hideous death. It was because of the manner of his death and the fact that his brace of pistols had disappeared that some voiced the opinion that you might have deserted rather than come into contact with the dreaded affliction." I sat there in stunned silence. One of the finest gentlemen, soldiers, and friends I have ever known had been taken from me without a word of farewell. Kismet could be so very unjust!

I spent the next few days coming to grips with the loss of my mentor. Because Mister Flemming had gone ashore on one of his self-imposed missions I had nothing of note to do. Consequently, I spent my time idling about **HMS Tigre**. Sir Sidney was kind enough to keep me there rather than dispatch me to **La Sangsue**, where Captain Creevey wanted to keep me locked up awaiting court-martial for desertion.[30] I was appreciative of Sir Sidney's sense of justice. He was a strange man, with many quirks that irritated the stodgy and complacent, and a total self-confidence which infuriated his less-able superiors, but he was as true a gentleman as I have ever met.

When at last Flemming returned he interviewed me in his usual energetic, no-nonsense manner. "What have you been up to in Cairo, MacHugh? Sir Sidney told me your tale of 'abduction' from the Gardens of The Butcher and of your service with Bonaparte. Captain Creevey would love to hear of the latter, so we'll say nothing more of that. No doubt he's bound and determined to make an example of you. Creevey's a hard and severe man -- fancies himself a disciplinarian. Being a civilian myself, I can say it. He is a tyrannical bastard who fears the men he commands and believes they will respect him only if they fear him. Stay away from that man, MacHugh. He has a great antagonism for Sir Sidney, and sees your complicated situation as a way of striking at our friend. Now, let's hear about your activities while in Cairo."

"Simply put, sir, I 'deserted' from the 32nd Half-Brigade and took back my old position as bodyguard to Sheik El-Bekri's daughter, Zenab. There I stayed until I was contacted by one of the Sheik's wives whom I know as 'Jasmine'. Sir, I am curious to know how she fits into your operations and how she knew about me. I had begun to fear I would be stranded there forever."

Flemming looked me in the face with a particularly penetrating gaze. At last he spoke slowly with an inflection that revealed his suspicions. "Fear, MacHugh? Fear? Yes, that would be a fear-filled situation -- stranded in the harem with el Bekri's beautiful but neglected wife, 'Jasmine', while guarding Bonaparte's lovely but abandoned mistress, Zenab. Lesser men would have broken under the strain," Flemming concluded dryly awaiting my reply. I realized he had left me an opening by treating my foolish comment almost as a joke. Had I revealed a 'give away'? I had not actually lied to him although I had avoided one vital fact.

Finally he spoke. "Is there something you have not told me, MacHugh?"

I stood there wondering what to say while shuffling back and forth from one foot to another -- an obvious give-away. With that realization the truth exploded from me in words I cannot begin to remember with any accuracy. When I had finished, my Secret Service mentor stared at me knowingly before asking severely but quietly, "MacHugh, what did I warn you was the single most dangerous aspect of our work?"

"Uh, the fair sex, sir," I answered quietly.

"So you remembered but ignored my warning. Well, didn't you, MacHugh?" When I did not reply he demanded quietly, "Answer me, MacHugh. Now!"

"I suppose so, sir"

"What do you mean, 'suppose so'? You damned well **know** you did. Don't you?" His voice was becoming louder.

"Yes, sir. What I meant was I didn't mean to ignore you, sir. It just happened. I don't know how. It is just that she is so beautiful, and was so lonely. I was just trying to help her and it seemed like good cover, sir, and then -- uh -- well -- ah." I had run out of excuses. I took a deep breath and confessed, "I **love** Zenab, sir."

He was glaring at me now. Quietly he almost snarled, "You silly young fool! That is no excuse. You 'love her'. Hah! We are dealing with thousands of lives, and whole empires are at stake. I did not train you and send you there to improve your love life! You are supposed to use your brain, not your damned fool heart." Although his voice had risen it was still under control. "Well, what about it, MacHugh?"

We stood facing one another. I had come to the 'Attention' position as the best way of dealing with the situation. I didn't know what to reply. Flemming turned and slowly stepped a few paces away as though regaining his composure. His shoulders rose then dropped as he took a deep breath. Turning, his face now calm, almost reassuring, he spoke quietly looking above and beyond me. "You were fortunate. We all were, MacHugh. No one died as a result of your passions. So relax, MacHugh, but let it be a lesson to you. Use your brain **not** your heart in these situations. Now stand yourself 'At Ease'. I am not your Colonel, not even your Corporal, so relax and pay attention -- **close** attention." I did so and he handed me a glass of excellent Scotch whiskey, although I suppose I was in no condition to judge it.

"You must understand, MacHugh, that our operation is very complicated. From certain ships here in the Mediterranean I send out branches which in turn sprout their own tendrils which penetrate the various places where we might learn something. In Cairo, for example, we have dozens of informants. Some are invariably accurate, and some rarely so. Although all of them supply information in exchange for cash, some request special favors -- political or personal -- others do it for revenge or spite. 'Jasmine', as you call her, is one of our best sources, situated as she is in the very house Bonaparte originally chose for his headquarters. She pays several of the servants there to report gossip, search for scraps of paper, and eavesdrop on the officers. She collects all this and sends it to the coast by courier. She arranged for her own couriers. However, over the last few months her entire courier network has vanished. Undoubtedly someone talked and they have all either been eliminated or have gone into hiding. For some reason of her own Jasmine did not wish to use you unless absolutely necessary." He paused as though expecting an explanation from me. I believed I knew the reason -- Zenab -- but said nothing. "Anyway," Flemming continued after a pause, "the destruction of her courier system forced her to employ you for this latest dispatch. 'Jasmine' is as wily as they come. It was she who initially apprised me of the Fourès affair. That is, I am sure, how she learned of you in the first place. Am I right?"

"Probably. I remember that Roustam Raza introduced us," I replied. "How does he fit into the network?"

"Roustam Raza?" mused Flemming. "I have never heard of him."

I told him all I knew of my friend and added, "He was recently given to Bonaparte as a gift by El-Bekri. He now serves as the General's valet. Roustam is in the best position of anyone to know what General Bonaparte is thinking."

"Yes, he does sound promising," mused Flemming. "I shall have you follow that possibility when the time comes. However, we have a pressing matter here at the moment. 'Jasmine's' message has fallen upon deaf ears, I'm afraid. Old Mustafa Pasha has wasted the valuable time she gave us by simply sitting on his duff instead of pressing along the coast or moving inland. Instead, he has fortified the peninsula there and has decided on a defensive stance. Meanwhile, Bonaparte has performed miracles of organization. Within four days of the Turkish landing he concentrated a striking force at Rahmanieh. So now Bonaparte doesn't even have to win a battle to defeat Mustafa Pasha. All he needs to do is starve our allies into submission. It passes my understanding why the Sublime Porte insists on appointing decrepit old codgers to command armies -- particularly against able young bounders like Bonaparte. Anyway, I have a job for you, MacHugh. There is one promising young fellow serving under Mustafa Pasha. We have our eye on him and expect great things in his future, so we have encouraged him to trust us and rely upon our backing. He is Mehemet Ali.

I am sending you as a sort of 'aide de camp', but your real job will be to act as a courier between him and Sir Sidney. It would also serve us all well if you can watch Mehemet Ali's back. He is not without enemies."

~

So it was that in my guise as a Mameluke I entered Aboukir by boat that very afternoon. The place had been strongly fortified. On the tip of the peninsula stood a tiny fort. Further inland three successive lines of entrenchments straddled the narrow neck of land. Although Bonaparte later claimed 18,000 Turks defended Aboukir, the fact is that less than half that number faced him.[31] Our gunboats and several Turkish vessels, were in position to provide supporting fire from either flank, so old Mustafa had at least chosen a strong defensive site.

I was directed forward to join Colonel Mehemet Ali. He was a handsome, athletic-looking Albanian of about thirty. When I presented myself and handed him my letter of introduction from 'Smit' Bey' -- as our allies called Sir Sidney -- he merely glared suspiciously at me. "A Mameluke?" he observed with surprise and obvious distaste.

"Not really, sir," I replied, hoping my Arabic sounded very English today. "I am -- sort of an Englishman -- really a Canadian, eh?" Seeing his questioning glare, I tried to explain. "I'm from America, sir, North America. We're -- uh -- sort of British." He was no longer interested. "Sir Sidney Smith wishes to place me at your service in the event that you desire to communicate with him."

"If you are an English officer, why have you chosen to appear as a Mameluke, that most useless class of parasites?"

As I was not an officer, did not consider myself to be English, and agreed with his assessment of the Mamelukes, there was not much I could say. "Convenience, sire."

"I humbly appreciate Smit' Bey's support and cherish his undeserved affection for me. You must convey my sincere appreciation to him at your earliest convenience," declared Mehemit Ali. He then smiled stiffly at me and strode off. Obviously, this fellow was a no-nonsense sort, who had scant interest in Levantine etiquette. I attempted to keep a suitable distance behind him, on the fringe of the group of toadies who followed him everywhere.

Mehemet Ali's troops were Albanians, like himself. They were good soldiers -- tough, well-disciplined infantry attached to their young commander. It never ceased to amaze me that the Ottoman Empire could, with complete confidence, field so many troops from such diverse races. Considering the brutality and corruption of the Ottoman administration, it is hard to comprehend the loyalty of these troops -- who were treated as cannon-fodder by the ancient commanders appointed by the Sublime Porte.

Although Mehemet Ali had suggested I return whence I had come, I remained by his side, or at least at his rear, all through the night. I kept my eyes and ears open to discover what Flemming and Smith had seen in him. I could understand how he would be distrusted by those above him who were less competent and lacked the charisma and dynamic energy of this man who 'suffered fools not at all'. In his own way, he was much like Sir Sidney Smith. I sensed that Mehemet Ali certainly did need someone to watch his back.

A SECRET OF THE SPHINX

Early the next morning, July 25, the French attacked. They threw themselves with sudden violence upon the Turks' first entrenchment. Once again the *grognards* were prepared to give their lives for the young butcher who promised them an abundance of Glory. I was astonished by their *élan*. The Turks met the savage assault courageously and put up a good resistance. Nevertheless, the surprise and the impetuosity of the French attack proved too much for them, and they gave ground as the morning wore on. I spent my time with Mehemet Ali as he rushed from one critical point to another. Many times he threw himself into the fray to encourage his men by example. On those occasions I felt it my duty to protect him without getting in his way. Consequently I have a very disjointed and episodic memory of the Battle of Aboukir.

At one point when Mehemet Ali waded into the front line, sword in hand, I was forced to throw myself upon a line of French bayonets armed with only my ancient scimitar and Colonel Phélippeaux's pistol. Swinging the huge curved blade with all my strength I managed to dash aside the bayonets that threatened him. I remember firing the pistol at a French officer who lunged at Mehemet Ali. I missed, but the fierce Albanian ran the Frenchman through himself. For me, most of the battle was spent watching my charge's broad back, or darting in to swing my scimitar wildly, but ineffectively. Such weapons are best handled by big men, I decided, and vowed to learn how to use the bayonet.

The battle raged for hours with the Turks falling back slowly though in excellent order. I now marvel at the endurance and stamina we all exhibited. It was about noon when disaster befell. General Murat's cavalry charged with such suddenness and such speed that they broke through the lines of Turkish infantry and attacked the small fort in our rear. As the Turks had no cavalry, it proved impossible to counter this blow, and the entire defense became a shambles. Within minutes all cohesion was lost. This enabled the French to break the defenders into small groups, some of whom fought to the death while others raced for the rear or waded out into the bay to swim for the ships. I later heard how our white-bearded old commander, Mustafa Pasha, was captured -- by Murat himself. The old Pasha fired his pistol at the dashing cavalry commander, wounding him in the jaw. Murat slashed at his opponent, knocking the pistol flying and taking off two fingers. The old man then surrendered and Murat bound his hand with his own handkerchief. The French never cease to amaze.

In the midst of the chaos Mehemet Ali stood defiantly trying to stem the collapse of his troops. If ever he needed a protector it was now! I was young and small, but I threw myself in front of him and yelled, "To the beach, sir! It's your only chance!" He glared at me, then looked about at the panic, and realized all was lost. Through the mob we pushed, reaching the beach at the same time as hundreds of others. Beyond, well out in the bay, several gunboats lay at anchor, firing over us at the victorious enemy. I shucked off most of my clothes and abandoned my huge scimitar but tied my multi-colored sash around my waist to secure Colonel Phélippeaux's pistol. Mehemet Ali had also stripped down, so together we splashed into the water crowded with hundreds of other naked wretches. I looked behind, and there, only a few yards distant, squads of French infantry were bringing up their muskets to aim at us -- an unarmed, naked mob in the surf. As their officer brought down his sword, I dove at my companion, and toppled him into the sea. When we came up, swirls of red mingled with the dirty green of the bay, and half a dozen men thrashed about in agony. Many others floated face-down around us. Mehemet Ali spared a second to stare at me then nod in acknowledgement before continuing on.

Together we splashed out on foot until it seemed best to swim. All the while shots rattled out from the shore, and every once in a while a cry of pain was the response. We would dive and swim out of sight as long as we could. It was strange to see the musket balls sizzling past us in the water until, losing their momentum, they dropped out of sight. Eventually we were beyond range of the muskets, and we tread water to look around. Hundreds of dead swimmers and waders lay bobbing in the water behind us, while the beach swarmed with French, taking pot shots or looting the bodies and the little piles of clothing. Around us pitifully few swimmers laboured on in silence Fortunately, the gunboats were not far away now, so we turned our attentions to them. In a few minutes Mehemet Ali and I were sitting panting on the deck of one. "You saved my life, little Mameluke," he gasped, looking me in the eye

～

The next few days were frustrating for all. Sir Sidney Smith strode up and down the deck of ***Tigre***, chafing at his inability to change the inevitable. In the fort at the tip of the peninsula, 2,500 Turks continued to resist. Their crowded state can be imagined when it is remembered that the same fort had been successfully defended by thirty-five Frenchmen for three days. The garrison was commanded by the son of Mustafa Pasha. The old pasha co-operated with Bonaparte in appealing to his son to surrender in exchange for a promise of safe passage in Turkish ships. The senior officers surrendered willingly, but the common soldiers were made of tougher stuff and they refused to surrender -- mindful, no doubt, of the Young Butcher's false promises to the garrisons of El Arish and Jaffa. Never have I felt more in sympathy with mutineers. They held out for a week under the worst imaginable conditions of hunger and thirst. Fortunately, our boats were able to evacuate a few of them in the dark. When on August 2nd they finally capitulated, only one thousand starving wretches staggered through the gates. Surprisingly, the French gave them food and water. The survivors then gorged themselves so ravenously that 400 died before they could be evacuated. I still suspect that their food had been poisoned. The Young Butcher was certainly not above such methods.

So ended the Battle of Aboukir, "one of the most beautiful battles I ever saw," as Bonaparte described it. Although the Ottoman army had lost thousands, the French had not gone unscathed, having over 200 killed and nearly a thousand wounded. Sir Sidney took some consolation in these figures. "A few more victories like Aboukir will serve to annihilate Bonaparte's Army of the Orient," he told us as he paced back and forth. Some cynical mathematician in the audience muttered, "Aye, all Old England has to do is sacrifice one-hundred-thousand more Turks, and we'll wipe the frog-eaters clean out of Egypt."

A day or two after the battle, two French officers came aboard **H.M.S. Tigre**. They had been sent by Bonaparte to arrange an exchange of prisoners with Sir Sidney. Always the courteous gentleman, Smith gave them several European newspapers to take back. When they left, Sir Sidney let it slip that he had once again been indulging in 'psychological warfare'. The newspapers were full of French disasters; their armies were being driven out of Germany and Italy by the Russians and Austrians while at home the Republic staggered from one crisis to another. Once this litany of disasters became known it was certain to rock French morale Sir Sidney assured us. It might even convince Bonaparte to throw in the towel rather than fight more futile battles for a defeated country. Just like Sir Sidney's last 'psychological' venture, the pamphlets at Acre, this too produced the opposite effect. Bonaparte was to use this ploy to trigger one of his most astounding -- and most cynical -- successes.

Chapter 12
"Farewells"
(Cairo to Alexandria, 12 - 23 August, 1799)

"Where did you go this time?" Zenab pouted. "I might as well recall Omar of Myra to be my guard, if you are going to slip away from me any time you feel like it."

"I only go when I am ordered to do so," I protested.

"Who orders you to leave my side?" she demanded. "It is *I* who give the orders. It is *I* who saved you when Aboonapart was searching all over Cairo for you."

"Although my heart is yours, darling Zenab, military duty commands me to be elsewhere on occasion."

"You still have not answered -- where were you?"

"I was at Aboukir." I didn't want to say more because I was not certain whether she would be rooting for her own people or for the love of her life, Aboonapart.

She turned white and gasped, "Oh, no! You might have been injured -- or even killed! To think that I might have lost both you and Aboonapart at the same time!"

Fortunately, this thought frightened her so much that she left off questioning me, and we made love to celebrate my return. As usual I returned to my chamber shortly after dawn to catch an hour or so of sleep. I was awakened early by one of the servants.

"You are to prepare yourself for duty tonight at the Feast of the Prophet," he advised me.

"But I have been assigned to guard the master's daughter. I do so every night."

"Never mind that. All of you loafers have been ordered to appear before the Master prior to the grand dinner for our guests. The Sheik wishes to impress them with his large retinue."

While I polished my weapons and boots I pondered this unexpected opportunity. "You are taking a leap forward in your duties, MacHugh," Mister Flemming had told me. "So far you have been a courier -- a very spasmodic courier -- but now you are going right to the heart of the business. Because of your friendship with this Raza fellow, I am entrusting you with the task of bringing him into our network." Now, on only my second day back in Cairo I would have an opportunity to carry out this assignment. The guest list for the banquet celebrating the Feast of the Prophet included Bonaparte. Probably his new Mameluke valet would accompany him. Consequently, that evening I stood under the arches behind the banqueters trying to look haughtily reassuring. There were about four dozen of us -- Mamelukes who had loafed about the palace for months on end bullying the servants, rogering the women -- and the boys -- while drinking the Sheik's wine-cellar bare.

Bonaparte was the honored guest. I was surprised, however, by the other guests; the Commandant of Aboukir, Mustafa Pasha, and several high-ranking officers captured there. They nudged one another with disbelief when Bonaparte joined in their rituals and prayers. Clad in his French General's uniform, his pale face and long stringy hair looking very European, he certainly had cheek trying to gull us with his performance as a devout Muslim. His arrogant confidence that he was hoodwinking everyone increased my hatred for this butcher. Standing in silence behind him throughout the charade was his new valet, Roustam Raza. Maybe that fact kept my urge to kill this multi-murderer under control.

I was watching El Bekri and Bonaparte exchange toasts when a voice spoke softly behind me. "Caspar, my friend, there sit two men who would be most interested to know how you spend each night." It was Omar of Myra. He smiled evilly then strode forward to join the guests, and surprised me by seating himself to the left of the master of the house. The Sheik smiled at him and placed a delicacy between Omar's lips. It was Omar of Myra who had ousted Roustam in Sheik El Bekri's affections!

For me the evening was both terrifying and disappointing. Terrifying because I feared what Omar of Myra might "let slip" to the Sheik, and disappointing because I never even got close to Roustam. In the days that followed I still found no opportunity to speak with him. He had moved his possessions to the apartments occupied by the General and his retinue at the palace of Elfi Bey. When I tried to visit him there I was brusquely told to "Bugger off!" by the French sentries guarding the place. Rumours circulated that the General was about to depart on a tour of inspection of all French forces in the Delta. That could take weeks, and as he would take his valet with him I would not be able to carry out my assignment for a long time. I was very young for this type of work and was anxious to please Mister Flemming. Not realizing how long these operations could take, I was feeling desperate. Imagine my relief then when Roustam turned up late one afternoon at my little chamber.

My normally swaggering friend looked almost distraught. "My life is ending, little brother," he wailed. "I am going far, far away with Aboonaparte. I shall never return to Cairo, my home of homes. Never again will I pleasure my true master, El-Bekri, whom I have loved above all others."

"Don't worry, Roustam! Aboonaparte is only going for a few weeks, visiting his soldiers in the Delta. You'll only be a few miles from Cairo."

"No! That is not true. Aboonapart has ordered me to pack all his possessions for a long journey. Then he ordered me to prepare myself to leave with him. We will depart at any moment. We are going to France!"

I was stunned. "That can't be, Roustam. The army is not ready to move. It will take weeks of preparation to evacuate such a large force. We would have heard if such a move was planned."

"Aboonapart is going alone! He said just he and I. No one else knows of his plans. No one except his humble guardian, Roustam Raza. I am afraid, Caspar. The French people are brutal and unpredictable -- except for you, my friend. They do not follow the true faith, and they wear silly clothes. I will be alone and reviled in France."

"Roustam, this may be a great opportunity for you. A great opportunity to serve all humanity *and* yourself. A group of wealthy and influential friends have asked me to speak to you about joining with us to rid Egypt of the French and bring back the old ways. If you can bring me information of Aboonapart's intentions, it will help your people throw off the yoke of the French invader. You will be free again." Looking back today, I am embarrassed to see how naïve I was then. Roustam was no more interested in that patriotic claptrap than in taking harpsichord lessons. But my next statement caught his interest. "Roustam, I have brought money with me -- a large sum -- which I am to pay you in exchange for the information you bring me. You will be able to help yourself while you help your people."

He looked up at that and his blubbering ceased. "These friends of yours? They are British? Mister -- ah -- Flemming is one of them?"

I regretted my indiscretion that afternoon on the Sphinx. "It doesn't matter, for even I don't know who they are. The point is that you will be well paid for reliable information -- but it must be reliable."

"Can you not pay me now? I have told you Aboonapart's intentions. You can pay me now, and I will bring you more intentions later and you can pay me again."

"No, I'm afraid I can't pay anything for this gossip you've brought me. After all, it's unbelievable! A general cannot just abandon his army and return home. If it were possible, everyone would know about it. This is Cairo!" Even at that age I had some common sense, so I added, "Besides, you don't think I'm such a fool as to keep the money here, do you? No, Roustam, when you have proof and details of Aboonaparte's plan, let me know, and I will meet you with the money."

"Little brother, I will meet you tomorrow at dusk in the small park across the street. I will bring the proof, and you will bring the money. We will both be very happy -- as will all enemies of the French."

~ A SECRET OF THE SPHINX ~

The next evening as dusk was about to descend I walked to the small park neighboring the Sheik's palace. Of course, I had been storing the treasure in my room -- there was nowhere else to hide it. I felt vulnerable carrying so much gold in a pouch tucked into my sash. For that reason I left while it was still light so I could chose my spot and keep an eye on all the entrances to the park. Also tucked into my sash, in the small of my back, was Colonel Phélippeaux's pistol. After a long wait the sounds of two horses announced an arrival. Within moments Roustam Raza appeared walking up the path towards me. It was now so dark that I could barely make out his features. "Have you come prepared to do business, my friend?" he asked. When I nodded he went on, "You wanted proof of Aboonapart's intentions. Tonight I will show you all the proof you will ever need. Come with me, Caspar. We ride to the house of Elfi Bey. We must go there quickly, for the General does not know I have slipped away."

Roustam had brought a horse for me, so we mounted and rode off. He would say nothing about the "proof". His interest was centered solely on the reward I had brought. On our arrival outside Elfi Bey's mansion we left the horses in a secluded spot and tiptoed through the grounds until we came to a garden traversed by paths. Hushed voices with the unmistakable overtones of excitement roused my interest. "Come with me," said Roustam, and I followed him deeper into the garden. "If you need proof that Aboonapart is leaving for France then look around you," he said with a self-satisfied smirk on his face. "Keep out of the way and keep your eyes and ears open, little brother, and you will see all the important people who are going to France with my master." I was about to slip away to see for myself when he put a hand on my arm and added, "Caspar, I don't want to lose the gold because you get caught spying here. Give it to me now."

I passed the pouch to him. Roustam hefted it before slipping it into his robes. "I must join my master now, but meet me beside the horses before you leave." Roustam had begun to accept Bonaparte, and was even beginning to relish the prospect of his new life in a new world.

My best opportunity to observe was to blend into the background so I found a silver platter bearing emptied glasses. Picking it up, I wandered obsequiously from group to group picking up empties. Never have my ears been more attuned for eavesdropping. The guests strolled nervously up and down the paths, afraid to move far from the center. On the main path, as befitting their rank, were the Generals gathered around a fountain. They were all young, I noted. The senior officers had not been included in this exclusive escape party. Ever since my first arrival in Cairo I had been trying to identify by name and face the most prominent of these Frenchmen who had turned the Levant upside-down. Now my efforts paid off for I recognized Bonaparte's closest friends. His Chief of Staff, Berthier, was there, as was Lannes, and Andréossy. I had seen them from afar, both at El Bekri's palace and before the walls of Acre. Now I could watch them up close. Berthier looked ill yet excited. Palace gossip had it that he suffered from a double affliction -- dysentery and love. Besotted as any teenager, the forty-six year-old Chief-of-Staff was almost delirious with the thought of seeing his lady-friend again.

Lannes was a great blonde bully of a fellow. Now he strode up and down in front of the others, his huge sword clanking against the bricks of the path. "I will most definitely have the truth from her," he fumed. "I ask you, comrades, if you were to receive letters like that wouldn't you demand satisfaction from the bastard?"[32]

"You had best ask friend Murat that question," laughed Andréossy. "He is more qualified to discuss such questions, I'm sure, and he is to join us later. Now, my dear fellow, I pray you put such thoughts aside and bask in the joy of returning once again to our beloved France. Aha! here is the final member of our little coterie!"

A hearty fellow in a green uniform strode into the light cast by the torches. It was Bessier, commander of Bonaparte's personal guard, "The Guides of the General".[33] I could see by the way they glared at each other that he and Lannes were not friends. "My escort of Guides is outside," announced Bessières. "We can leave anytime the General wishes."

"Thank you, Bessières," came a familiar strident voice from behind me, and I'm sure the hairs on the nape of my neck stood on end. I shuffled off to the side as Bonaparte himself stepped onto the flagstones surrounding the fountain. "I trust none of you were inconvenienced by my sudden summons. If so, you are welcome to remain in Egypt," he added with heavy-handed irony. "As you know from the newspapers given us by that damned Smith fellow, the situation in France is grave. Anarchy threatens, with disasters multiplying on our borders. The conquests I made for our beloved France are being lost by incompetent and corrupt politicians and their so-called generals. I must answer my nation's call to return. It is my destiny to succor France and restore to her the glory she deserves. I am but her humble servant, so I must return."

"My General," said Bessières, "we are honored that you have chosen us to accompany you in your quest. My life is at your command. Send me where you will." Not to be outdone, the others enthusiastically avowed their loyalty with a babble of similar phrases.

Their platitudes were interrupted by the sounds of a carriage arriving. I worked my way over to view the new arrivals, three civilians[34] clustered together as though afraid of missing something. Two of them, Monge and Berthollet, I recognized at once. Every soldier who had been to Syria recognized them. The two inseparable savants had served the General on so many commissions and had been so many places with the Army that the *grognards* laughingly lumped them together as "Mongeollet".

"Imagine what the members at the Institute are saying at this moment!" exclaimed Monge.

"I hope the General doesn't discover that you let the cat out of the bag," laughed Berthollet. "You were positively incoherent with excitement -- babbling like an idiot he was, Denon," he continued to the third member of the party. I moved slowly, busily wiping the glasses as I listened. "You know, don't you, Monge, that Parseval rushed upstairs to pack his bags once you blabbed? Can you believe it, Denon? My partner here denied all by announcing, 'We are **not** going to France, but if we are, I didn't know about it till noon.' That should certainly quell the rumors, don't you think?"[35] While they laughed I moved on, thankful that in the excitement of the moment, no one seemed concerned by the presence of a well-armed Mameluke 'servant'.

Next I found Bonaparte's four young aides-de-camp chatting coolly, endeavoring to conceal their excitement. Bonaparte's step-son, Eugene Beauharnais, I remembered well from Jaffa.

"Word arrived today from Admiral Ganteaume that the Anglo-Turkish fleet has set sail, and is leaving Egyptian waters," he was telling them. "The next few days will see our squadrons in control of the sea-lanes to France."

"I thank my lucky star that your step-father has decided to include us," said one, poking playfully at Beauharnais. "Just think, fellows, in a few minutes we'll be on our way home!"

I was thunderstruck. They were leaving tonight! Somehow I must get word to Flemming. If I moved quickly, we might capture Bonaparte at sea, leaving the Army of the Orient leaderless and at our mercy! I must get out of here. Abandoning my tray, I looked for a way out. A secluded path disappeared into darkness on my right so I headed down it. Rounding a corner I almost crashed into a tiny hussar, only saving the other from a nasty fall by throwing my arms out to catch him. It was her -- 'Bellilotte' Fourès!

"Oh!" she gasped in astonishment as I held her in my arms. Her sweet baby face was only inches from mine -- for the second time.

"Pardon! I'm so sorry," I blurted out in French. She must have realized it was an accident, for once her surprise passed, she did not look frightened. Realizing I was still holding her waist, I released my grip and stepped back embarrassed, and bowed. She smiled coquettishly at first, then her ivory features took on a puzzled expression.

"I know you from somewhere," she said. "And you speak French -- but you're a Mameluke. You wouldn't know French."

Footsteps sounded behind me on the darkened pathway. I stepped past Bellilotte, leaving her puzzling over our second meeting. Soon she would recall the first, dancing a similar duet, but nude, in her chamber. I ducked into the shrubbery beside the path. My intention was to slip out of sight in case she sounded the alarm. However, I immediately ran into a high wall. There was no quiet way over, so I crouched there hardly daring to breathe.

"Bellilotte?" came Bonaparte's shrill voice. "I thought I heard you. You were with someone."

"Yes, dearest. A strange Mameluke ran into me. He seemed familiar, and he spoke French, but he meant no harm. He seems to have disappeared somewhere. I --"

"Never mind him, my sweetheart," broke in Bonaparte. "I have something to tell you." I heard the rustle of clothing, and my experienced ear detected hints of kisses. "I must leave you for some time, my sweet Bellilotte. Duty calls. Even more than to you, my dearest, am I the slave to Duty. But you understand, I know, and will think of me as I shall think of you every moment I am away. I shall return for you soon, my lovely Bellilotte."

After a few moments I peeped out of the shrubbery. Bellilotte Fourès stood alone sobbing. She knew he was lying.

Anger filled my heart. I had once again missed the chance to kill the loathsome butcher of thousands of innocent children and gallant men. Now he would continue his murderous career, and millions more would die to satisfy his vanity. I felt like sobbing alongside his abandoned mistress.

I returned to the horses. Roustam was already there. "You took a long time, my friend. Aboonapart will be angry if I do not come when he calls, so I must go now. But first, something very important for you and your friends. A man of the greatest importance to your cause is waiting even now for you at the Sphinx. If you wish to meet one who holds Upper Egypt in the palm of his hand, you must ride there tonight to meet him. Murad Bey awaits you at the Great Sphinx. Go, my little brother. Allah go with you." Roustam embraced me and was gone.

∼

The Pyramids bathed in moonlight are unforgettable. A strange pale luster is reflected by their slanting surfaces, and they glow, visible for miles despite the blackness. Nestled below the largest of the three I would find my old friend, the Sphinx. I had been forced to pay exorbitantly to obtain passage across the Nile at this time of night, and had then put Roustam's stallion to a gallop despite the unsure footing. Now we slowed to a walk as we crested the last dune before the great stone guardian of the desert. A hush descended.

The Sphinx's misshapen face gazed benevolently at me, its features made more startling by the intense white moon-glow. I had hoped to see my snowy owl floating above it, but only the eerie moon and a myriad of stars stared at me. Strangely enough, those lowest on the western horizon seemed obscured by a haze. I stared again at the Sphinx, its face as inscrutable as ever, yet this time I sensed a warning. Searching the landscape from left to right, I found nothing to cause alarm, and so continued on in the strange silence, halting beneath the Sphinx's right ear.

My friend's eyes seemed closed in peaceful sleep -- but? Was he trying to tell me something? Should I wake him? Cautiously I reached out to touch his neck. Suddenly a streak of white yanked my eyes upward! My Spirit Owl -- **and** behind him a ghostly scimitar flashing downwards at my head! I threw myself from my mount, rolled sideways, and sprang to my feet for behind the scimitar I had see a face. But no! It couldn't be!

Terrified, my stallion galloped off leaving me to face a silent black figure. He stood in the familiar attack stance -- feet apart, a huge scimitar in each hand.

"Murad Bey?" I asked hopefully.

"Hah! You are such a fool to believe Roustam Raza." It was Omar of Myra!

"Where is Murad Bey?" I asked.

"Who knows, *infidel*? It was **me** you were sent to meet. Sent here to die, duped into my trap by Roustam Raza. You are not the first he has betrayed. You now know, too late, Caspar, that it was not wise to take Zenab from me. But now *I* have the Sheik -- and tonight I shall have Zenab as well."

He leapt at me suddenly, arms churning like a windmill, his scimitars flashing in the moonlight. From my training with Roustam I knew how difficult it was to defend against such an onslaught. Desperately I parried with my lone scimitar, but I was already losing ground. I had to get one of the scimitars away from him. The sand caused him to lose his footing momentarily, and gave me my chance. I turned and darted towards the rear of the Sphinx. There was Omar's horse! I leapt upon its back and stood upon the saddle. Gripping my scimitar in my teeth, I used both hands to pull myself up, atop my enigmatic friend. As I scrambled up, Omar swung a scimitar, missing my foot by inches. He stood beside the horse, hurling insults at me. I laughed. A strange hum filled the air undetected while we had battled beneath the Sphinx's head.

ᔓ A SECRET OF THE SPHINX ᔔ

In moments he was scrambling up the rocky flank the same way I had. I refrained from attacking him as he reached the top, a foolish gallantry I would soon regret. Omar rose to his feet. He now held a pistol in each hand. Both were pointed at my heart. "You ran from me," he snarled. The ever-increasing hum of the wind forced him to finish by shouting, "I need not slay you like a man. I shall shoot you like the dog you are."

Behind my attacker fewer stars dotted the sky, and something barely visible swirled in the black heavens. His figure was somehow growing dim, and his voice seemed distant. "You came to my master's house to spy," declared Omar. "I have watched you, Caspar. You are not a Mameluke, nor are you of the True Faith. I have listened as Zenab has instructed you. You have come amongst us to steal the secrets of the house of El-Bekri."

Behind him, more and more stars were vanishing. A golden cloud swirled in eddies, each of the millions of tiny grains of sand reflecting the moon's pale light as it was lifted from the desert floor by the oncoming wind. The moaning had grown, but Omar, in his euphoria, did not notice. He simply spoke louder. "I know about you and my master's wife, the mother of Zenab," he boasted. "I too was pleasuring myself with one of my master's wives when you last visited her. I watched you leave her bed, and I saw you ride off immediately afterwards." The wind began to tug gently at our robes, and behind him the golden cloud now covered half the sky. Oblivious, Omar raised his voice above the wind. "After I have killed you I shall return and slay your accomplice My master has been badly served by a wife who spies for the hated French."

I had to gain time, so I said, "You did very well, Omar You almost got it right."

"Almost?" he queried, looking unsure for the first time. "What do you mean, 'almost'?"

"We do not spy for the French," I replied contemptuously. "I despise the French more than you do. They killed my mother when I was a child." I had to stall for time, so I smiled conspiratorially.

"But you do spy! I have heard you, infidel!" The sky behind him was disappearing rapidly. By now we were forced to shout to be heard.

"I spy *against* the French."

"How can that be, when your accomplice, Roustam Raza, is one of them?" The wind had risen to a threatening roar. It whipped our clothes about us. The swirling curtain of sand was almost upon us. The world darkened as the moon vanished.

"Roustam was not with the French. He worked for us," I shouted above the wind. "Now El-Bekri has given him to Aboonapart."

"Roustam has tricked you, infidel," he shouted. "Tonight your friend betrayed you again. He promised he would send you here so I could kill you. Before that he murdered seven of your accomplices -- Ahh!" The first swirls of sand lashed him a split second before it enveloped me. He winced and turned his head to avoid the stinging sand. I threw myself flat on my back, at the same time wrenching my pistol from the back of my sash. Squinting, I could make him out, a grotesque silhouette whose robes billowed in the wind. I aimed for what I hoped was his head. Two flashes lashed out and his pistols roared above the storm. I pulled my trigger and Omar of Myra toppled backwards out of sight.

A SECRET OF THE SPHINX

∼

I slipped into the silky warmth of the bed and slipped an arm about her. Instantly she rolled into my arms, her laughing eyes and her groping hands evidence of her delight. "Caspar!" she whispered.

"I have come to warn you, Jasmine," I murmured in her ear. I felt her tense momentarily. "I have just killed Omar of Myra. He knew about us, about our work, although he believed we worked for the French."

"Then he is no longer a danger to us. Let us make love, Caspar, my young lion."

"That is not all," I persisted. "Aboonapart has already left Cairo. He is returning to France -- tonight. We must warn Flemming to recall the fleet and capture him."

She began to kiss and nibble at my lips, interspersing her reply with sounds of rapture -- which I trust were genuine. "It is too late. Mister Flemming has sailed with them. They go to Cyprus for supplies. Let Aboonapart also go. He can do us no more harm. The French will either go home or will rule with greater wisdom after he has gone. Whichever happens, Egypt will return to the way it was. Now let you and I celebrate our victory by making love, my young blue-eyed lion."

∼

Next morning I entered the headquarters of "*l'Institute d'Egypte*", the scientific and artistic commission invented by Bonaparte to accompany the Army of the Orient on its expedition of pillage and slaughter. I was looking for one of the members listed on the sign at the entrance --"François Auguste

THE MacHUGH MEMOIRS ~ (1798 - 1801)

Parseval-Grandmaison: Poet". He must be the 'Parseval' I had heard Berthollet mention last night. Claiming to bear an important message for the savant from Berthollet, I passed the sentries and entered the grand lobby to enquire as to his whereabouts.

After the hot sun the vaulted hall seemed dark and cool. There seemed no one about, but in the center of the vast hall lay a large black stone. Curious, I stepped over and studied it. Although two of the corners had been broken off, it appeared to be some sort of plaque or tablet, for it was covered with strange script neatly set out in lines similar to those on a monument. There were three sections, but none made any sense to me. The top third was covered in "*hieroglyphs*"-- a pictograph type of writing I had seen on the ancient ruins around the city. These were mostly squiggly lines with a few recognizable pictures such as birds which reminded me of the drawings carved in stone at "It is Written" on the Opaque River.[36]

"Are you interested in antiquities?" came a voice from behind. I turned to discover a cheery but somewhat bleary-eyed face peering at me over the back of a sofa. "You Mamelukes are usually more interested in destruction than in preservation," he laughed. "Or is it plunder you're after? This is a great treasure for mankind, but not worth your bother to drag away. You couldn't get a worn *centime* for it even if you could move it." Seeing my curiosity had been piqued, he went on. "Although not one of my prime interests -- I am more interested in the study of the fascinating fish and reptiles you have in the Nile -- I will explain what little I know. This stone was found a few weeks ago in the ruins of the walls of a fort near Rosetta. It's almost two-thousand years old. It was inscribed by someone very important in his time, a Greek -- sort of a king, but even more powerful. It holds the secrets of antiquity, the great wisdom of the ages -- ***IF*** one could but unlock its mysteries. Someday you will tell your grandchildren that you saw it -- even touched it. This stone will become famous -- as will those who unlock its mysteries."

I reached out and gingerly touched a carved figure that looked like a witch on a broom. It felt cold, but there was no sense of communion such as I had when touching the Sphinx. Disappointed, I recalled the purpose of my visit. "Is Monsieur Parseval-Grandmaison here? I have a vital message to deliver to him in person."

"Yes. He is here. He missed the 'escape carriage' last night -- but only by a moment or two," laughed the scientist rising from the sofa. "It's the first time I've seen friend Parseval show signs of life since we arrived in this god-forsaken place," he laughed. "I believe 'The Bard of the Nile' might still be in his room -- contemplating suicide, I wouldn't be surprised."

My guide, whom I now believe was Geoffroy Saint-Hillaire, the great zoologist, directed me to Parseval's room. There I found the poet passed out on his bed. I slapped his face and dragged him off the bed. His bags were packed I noted with satisfaction. Only an open brandy bottle remained unpacked. He still wore the clothes he must have been wearing last night.

"Monseiur, do you wish to return to France?" I asked.

"Oh yes! Most certainly!" he wailed.

"Then get off your ass and come with me. You must move quickly though." It took a few minutes, but I coaxed him into a carriage, and we were off to Bulaq. En route I endeavored to jack up the little man's spirit and confidence. He was really a ridiculous, insignificant twerp -- the first poet I had ever met, and not a great introduction to the tribe. I began to feel I had made a mistake in choosing Parseval. My plan had been to catch up to Bonaparte myself. Using Parseval as my stalking horse, I would pose as his servant until I got close enough to Bonaparte to take action.

At Bulaq, using Parseval's money, I rented a boat to take us to Alexandria. This would be a journey of several days duration, but Bonaparte's entourage would take just as long, so we should arrive only a few hours behind them. During the voyage, Parseval recited a litany of complaints about Egypt, about other members of the Institute, but above all against his patron, General Bonaparte. "The General does not seem to realize that poetry requires inspiration. One cannot simply compose poetry of great merit by being bullied," he complained. "Shortly after we arrived in this miserable place I composed an excellent cantata for him for the Festival of the New Year, but the General was never satisfied. Can you believe he insisted that I do the work of a mere literary laborer by editing his newspaper, *Le Courrier de l'Egypte,* for him. I am a gifted poet blessed by the muses, as you no doubt know. I respond only to **their** inspiration. I do not simply turn a spigot labeled 'poetry' to pour out something worthy of my name. Yet that arrogant little man insisted that I write great epic poems lauding his person and commemorating his victories. Young man, did you see that magnificent stone tablet in the Institute, with its centuries of secrets waiting to be unlocked? Now **that** is something to inspire an ode! Ah, yes!" and he paused to ponder the possibilities for an hour or so.

"Can you believe, my friend," Parseval commented later, "that General Bonaparte was positively rude when he learned that I had commenced as my major work, an epic praising the capture of Acre by Richard the Lion-Hearted?"

My heart plummeted when I heard this last complaint. It would be difficult to find a literary effort more likely to offend the loser of the latest siege of Acre than a poem by a Frenchman praising an English king who had succeeded at the same task. It would take a huge heap of ingenuity on my part to place this poet close enough to the Young Butcher to enable me to carry out my plan -- which I can find no delicate way to describe. I intended to assassinate Napoleon Bonaparte.

~

Our boat pulled alongside **La Murion** as the sun was about to peek above the horizon. Sailors were moving about the deck, preparing for the wind which was freshening with the dawn. Obviously this vessel was about to set sail. We had tracked down Bonaparte just in time.

"Ahoy! Ahoy!" called Parseval. He was so excited I feared he might topple overboard in his agitation. We had followed our prey to Alexandria, only to discover last night that Bonaparte had already boarded **La Murion** which sat anchored offshore ready to start the dash to France. The 'Bard of the Nile' had been on the verge of tears at the thought that he had missed the boat. But I did not give up. Using his dwindling supply of cash it had taken me some time to locate and rent a small craft to take us out to the French man o' war. Now that we had arrived alongside I would have to leave it to my desperate accomplice to get us aboard.

"Ahoy! Monge! Berthollet! It is I, Parseval-Grandmaison. Let down a ladder!"

This announcement caused some consternation aboard and much shouting and cursing. Eventually Berthollet appeared above us. "Parseval! What on earth are you doing here?"

"Thank God I'm not too late! Berthollet, old friend, get someone to lower a ladder. I was afraid I would not make it in time!"

"What is all the racket about?" came a high-pitched screech I recognized. "Why not fire a broadside and set off illuminations? Silence, you damned fools!" Bonaparte raged.

"Sir, it is Parseval-Grandmaison," announced Berthollet. "He has just drawn alongside. He requests permission to come aboard, sir."

"That wretched little scribbler?" hissed the General in disbelief. "Tell him to bugger off! Tell him to scurry back to Cairo and finish his ode to that damned Englishman."

"But sire," protested Parseval, "*Richard Coeur de Lion* was only born in England -- he was of French blood!"

"Sir," came Monge's voice, "I'm sure Parseval-Grandmaison is already working on an epic in praise of your many victories. If you bring him aboard you will enable him to complete it, to the great enhancement of France and French letters."

"He can rot on the shore for all I care for his damned bloody scribbling!"

It was Berthollet's turn. "Sir, you could dictate your memoirs to him. He could produce the great epic of your life in quatrain. Think of the effect it would have when you step ashore!"

"Gentlemen, you speak nonsense!" was the Young Butcher's reply, but in a more controlled voice than previously. "However, let it not be said by my critics that I took only the best with me when I left Egypt. Let them admit that I spared my successor the whining and sniveling of 'Parseval, Poet of the Pyramids', 'The Ass of Acre'." The voice was rising in both pitch and anger. "Tell the damned plagiarizing little piss-peddler that he can scurry aboard, but you two are to ensure that I **never** set eyes upon him! If I so much as hear a word about the useless little shit-merchant I'll have him thrown overboard tied to an anchor!"

Consequently, a ladder was thrown down, and unobtrusively I set about carrying up Parseval's luggage while he gushed his appreciation to his colleagues.[37] Once I set foot on deck, I looked about for Bonaparte, but he had taken himself off somewhere. I made to slip away from this crowded section of the deck, when I felt something jab me in the back. It was a pistol!

"Caspar, you should not have come here," came Roustam Raza's hushed voice. "Now walk quietly this way without attracting attention from the *moallem*."

I did as ordered, and presently we faced one another alone on the stern. Within the folds of his robes, hidden from any passers-by he held a pistol pointed at my guts. "Why did you come here, little brother?"

"I hoped to confront Bonaparte."

"You wish to kill him. No one will **ever** murder my master. You know, little brother, from this day on, I, Roustam Raza, am his bodyguard."

"I also wanted to speak to you, Roustam, my 'friend' -- about my meeting with 'Murad Bey'. Why did you send me to the Sphinx to be murdered?"

"I sent you to kill Omar of Myra. -- which I see you have done. I knew you would kill him, little brother. Omar of Myra was a fool to trust me after he stole my place in the affections of

El-Bekri. But Allah has smiled upon Roustam after all. This new master is the greatest man I have ever encountered. Aboonapart will be master of the world one day, and I, Roustam Raza, will stand behind him. I share his Kismet."

"How long have you served Aboonapart, brother?" I asked. "You once worked for your own people. I remember when I met you at the Sphinx and you guided me back to the coast."

"You mean to Mister Flemming and Smit' Bey." He paused as though deciding how much to reveal. "You see, Caspar, I worked for Aboonapart even before I met you. Aboonapart was paying me to watch over his woman and tell him what I heard in El-Bekri's house of spies. My master's harem was a comfortable place to learn more than love-making, was it not, little brother? When you and I first met I thought we might become lovers, but when you rejected me I began to think of you as my little brother. I have never had a family, so you became my family. That is why I never told Aboonapart about you and his Zenab.

"One day Aboonapart ordered me to find the person who carried messages from the harem to the British. Naturally, I did as Aboonaparte asked. It cost six men their lives, but eventually I found him, the seventh man. But this man would not talk -- at first. However, after he had been castrated and blinded he pleaded for death. I offered to grant his wish -- but only if he would tell me who the secret courier was. He accepted my offer, but claimed he did not know the courier, only that he would be wearing a sash that matched his. After I cut off his hand, I realized he was telling the truth, so I took his sash and the message, which no longer mattered, and next day went to meet the courier at the Sphinx in his stead. It was you I met, there, Caspar.

~ A SECRET OF THE SPHINX ~

"I could not kill my little brother, so I spared you, but I followed the directions given me by the dead man. I delivered you to the boat that took you out to your master. I have since learned he is called 'Mister Flemming'. If you search diligently you will find that every one of those who assisted El Bekri's traitor-wife have now vanished from the face of the earth. Everyone but you, Caspar. I have not told Aboonapart about you."

I was appalled by his matter-of-fact story, and stood in shocked silence. He continued in an offended tone of voice. "One day while you were away -- in Acre, you say -- I was betrayed to Sheik El-Bekri by someone. My master dared not kill me and risk the wrath of Aboonapart. So he gave me to Aboonapart upon his return. It was El Bekri's little jest."

"So you were just pretending when you mourned about being given to Aboonapart," I said. "You were lying to me, my brother."

"Only a tiny bit, Caspar. In my heart I did not want to go from El Bekri. He had betrayed me, but I would never have betrayed him. Aboonapart could have continued paying me, and I would have given him what he wanted, but I would never have betrayed my master." I was amazed to detect a catch in his voice.

He looked me in the eye before changing the subject. "Caspar, the sun is rising. In moments Aboonaparte and I will sail away to France. You are a spy who has set out to kill my new master. I must be loyal to him, my little brother. Turn around now and face the sea and the rising sun."

I did so, and heard him draw his razor-sharp scimitar. "You understand my dilemma, Caspar; to remain loyal to my master I must kill you. It would not be seemly to disturb Aboonapart by firing a gun so I must employ my scimitar. I shall count to three like the Frenchmen do. When I reach the fatal number I must swing my scimitar and lop off your head, little brother. I hope you understand." There followed a pregnant silence. "**One** --," he said slowly and paused. "***two*** --."

I dove into the warm green water as the sun peeped over the rim of the world.

PART III

"THE PIPER"

Egypt March - July, 1801

"A Piper of the 79th" (a self-portrait?)

CHAPTER 13
"A Kilted Red-Coat"
(Aboukir to Ramanieh, 8 March - 9 May, 1801)

"Piper MacHugh, strike up a tune to take us ashore," ordered Captain Carnegie. I stood up in the longboat and blew into the pipe-bag under my arm. A quick twist on the drones to re-tune them, and I started "***We Will Take the Good Old Way***". This tune was popular with the men; the oldest veterans remembered marching off to it the day the regiment was raised; those who were Londoners claimed it sounded a bit like "***London Bridge***". It is such an easy tune that I could gawk over the sailors' heads as they rowed us ashore. Our men looked very smart, fifty per boat sitting packed between the rowers, erect and rigid as statues trying to see what was happening on the beach ahead. Finally! I was about to land in Egypt again after an absence of a year and a half!

A lot had happened in those eighteen months -- none of it very significant to the world. The whole period now seems a blur of disappointments and boring routine. Watching the low waves roll up the sandy beach I remembered my departure from Egypt. I had presented myself to Mister Flemming aboard ***H.M.S. La Sangsue***. I mentioned the large black tablet I had seen at the Institute in Cairo and was surprised when

Flemming displayed great interest. "I've heard of this 'Rosetta Stone' as they are calling it. Seems the French scholars are in a big stir because one of the inscriptions is in Greek, and they claim it will become the key to translating the other two inscriptions. Of course, even the so-called 'experts' are unable to read the 'hieroglyphs' inscribed on the ancient sites. I've seen a few myself, and they are a mystery. Anyway, after Bonaparte's return to France the newspapers carried articles about 'the stone to unlock the secrets of the ages'."

At that moment I had encountered Captain Creevey for the first time when he interrupted our conference to have me put in irons as a deserter. Mister Flemming could do nothing about it, and while he was away trying to find Sir Sidney to free me, **La Sangsue** sailed for Gibraltar. That was a bad period for me. I was a prisoner again, and the injustice of it raged within me. Only two things kept me from going mad -- my Vision and my faith in Sir Sidney Smith. The latter came to my rescue when we eventually landed in Gibraltar.

I was paraded before Captain Creevey. I remember his thin white face turning red with rage as he informed me, "I have here an order signed by Captain Smith, with instructions to turn you over to the Army for return to the 79th Cameronian Volunteers. It is lucky for you, MacHugh, that you have a protector with the influence of Smith. Why he should wish to stand between you and the justice you deserve is beyond me, but let me assure you, MacHugh, I shall investigate your case and your background. I have not finished with you. You are a deserter, a renegade of the worst kind. You stole from your officer, Colonel Phélippeaux; you then deserted him in his hour of greatest need, and Lord knows what you have done since. But I shall find out, and shall bring you to justice. Don't think you have gotten away with anything." Consequently, fifteen months after enlisting, I sailed off to meet my regiment for the first time. I joined the 79th in December of 1799 at Chelmsford in Essex.

So thankful was I for my release from **La Sangsue's** brig that I threw myself whole-heartedly into my lowly role as a Private Soldier in the British Army. Hours of daily drill and Spartan living, being hounded from dawn till dusk by Sergeants, Corporals and senior Privates, seemed luxury after life aboard **La Sangsue**. I seldom cast my thoughts back to the palace of Sheik El-Bekri, nor to my youth on the Great Plains. Such thoughts would only lead to disenchantment and longings that would make my new life a burden. My Vision had told me I would return to the Great Plains and my Blackfoot family someday. The certainty I felt made everything bearable -- even challenging. I must admit that I actually grew to enjoy a soldier's life.

The 79th Cameron Highlanders became my new family. It was a good regiment -- the best, I thought. Experience has since shown me that every soldier thinks his regiment to be the best. It would be a pretty poor outfit that didn't consider itself elite. We were a strange mixture. Although officially "Highlanders", fewer than half of the men actually came from Scotland, and not all of those were from the Highlands. Lieutenant-Colonel Cameron had raised the regiment in Stirling in 1793 as a genuine Highland regiment, although the English pen-pushers in London mistakenly authorized the title to be "The 79th Cameronian Volunteers". Officially that was still our name, but we all refused to use it, because the "Cameronians" were an obscure sect of puritanical Presbyterians who had nothing to do with the Highlands. Besides, they already had a regiment of their own, the 26th Cameronians, recruited in the Lowlands. The 79th had been shipped off to the West Indies there to be so decimated by disease that it had been ordered home -- but not before all the remaining Privates were transferred to the 42nd Black Watch despite a royal order prohibiting it. When the skeleton of the 79th arrived in England Colonel Cameron had put up a tremendous battle with the powers-that-be just to keep his regiment alive. With great reluctance, the officials authorized him to recruit the 79th up to strength -- but only if it could be completed within weeks. Due to this restriction most of the 79th's recruiting was done in London. That was how I happened to enlist.

As a piper I enjoyed special privileges -- slightly higher pay, and prestige as an 'artiste'. There was also always the chance of earning a few pennies playing for weddings or other special events. The only fly in my ointment was the presence of Taddy Fromm. The day after I had joined the regiment in Chelmsford "Taddy Smith", as he now called himself, had turned up to make a claim against our friendship. Taddy's military career had not been distinguished. He had soon become known by his comrades as a coward and a shirker. His theft of several pennies from a mate had resulted in a flogging -- one of very few administered in the 79th. Private Smith was not cut out to be a soldier, particularly in a highland regiment where moral standards were more demanding than in an ordinary line regiment. As Corporal McAnsh put it, "Private Smith, ye'll never shite a soldier's turd." Yet Taddy had eventually behaved acceptably in action at Egmont-op-Zee in Holland while I had been 'guarding' Zenab and swimming with Mehemet Ali.[38] Now the green recruits with whom I had enlisted were veterans, and Taddy liked to lord it over me. I had not been able to shake him loose, so it looked like he and I were still 'mates'.

Though only a lowly "lobster", I maintained my interest in events in the Levant. My old friend, Sir Sidney Smith, continued to be alternately lionized and vilified by the press and the politicians. I was one of very few Privates who read the broadsheets, and although you could never really trust them, you could, if you kept your own wits and experience in mind, understand what was happening in the Mediterranean. I already knew Sir Sidney had earned tremendous respect from the various races and factions of Egypt and Syria. He was trusted as a sort of "honest broker", and went where few others would dare to venture. In January, 1800, at the insistence of Lord Elgin, the new Ambassador in Constantinople, Smith had brokered a peace arrangement, the Convention of El Arish, between the Grand Vizier and General Kléber, Bonaparte's reluctant successor. Our Government had repudiated the agreement almost immediately, because of the concerns of our European allies who feared a French army returning from exile. Abandoning Sir Sidney to face the music, Lord Elgin had back-tracked adroitly and had convinced the Grand Vizier to attack the French for an easy victory. The results were the virtual destruction of the Grand Vizier's enormous army at the Battle of Heliopolis and a bloody insurrection in Cairo which the French put down with equal ease. The final reckoning would come a year later in 1801 but only after thousands of Turks, Egyptians, French, and Britons were killed or mutilated, just to sign an identical treaty. But I am getting ahead of myself.

Our regiment -- in fact, our entire army -- was now pretty demoralized. Although we considered ourselves good soldiers in an excellent regiment, we were convinced we were led by fools and cowards. We despaired of ever winning a victory. For instance, the 79th had been part of the embarrassing fiasco called "The Ferrol Expedition". On August 25th, 1800, we had been disembarked in Spain to capture the port of Ferrol. The next day we were re-embarked after having done absolutely nothing. For our next humiliation we were shipped to Cadiz, but that landing was cancelled for some reason, and back to Gib we sailed. Since then we had been hit by a hurricane and driven out into the Atlantic for weeks before being taken on a Mediterranean cruise, leaving sick soldiers in every port of call. It was a miracle that all of us were not sick. Maggoty biscuits, salt pork, and foul water we shared with the sailors -- but seasickness, the inactivity and grinding wretchedness of being kept below decks as "live lumber" for days on end drenched in bilge water had made our lives absolute hell. Thank God for the pipes! Because of them I got on deck almost daily to provide entertainment and accompaniment for the tars while they worked, and even managed to churn out a few tunes below-decks for our own fellows, although it was not necessarily enjoyed by all the seasick Cockneys.

In Malta we added to our Highland flavor when about three hundred true highlanders from the Fencibles joined us. Our last port of call was Marmorice Bay in Turkey. Here, far from the known world, we practiced landing on a defended coast. It was great to set foot once again upon Mother Earth! There were a lot of mishaps and cock-ups, but we persevered for weeks along with many other regiments. Rumour had it that we were destined for Egypt to take on Bonaparte's old Army of the Orient now under its third commander.[39] The soldiers' spirits began to soar, but for us Camerons there was a fly in our beer.

The commander of our expedition was General Sir Ralph Abercrombie, a relative of Lord Dundas, Minister of The War Office.[40] To the 79th he was the most hated man in the Army. It had been Abercrombie who had ordered the destruction of the regiment in the West Indies by forcibly transferring the 79th's remaining Privates to the 42nd despite the written guarantee of the King. Being commanded by Abercrombie, Scot though he was, filled the 79th with rage.

Two months later we left Marmorice Bay, bound for another landing -- 'the Delta', according to rumour. After several stormy days at sea, we arrived in Aboukir Bay, having had some of our transports scattered in the storm. Due to the weather and tides we sat at anchor seven miles off shore where the French could watch us. That gave them a week to reinforce their defenses before we landed. It looked like Ferrol and Cadiz all over again -- but with slaughter as the finale.

At last, shortly after dawn this morning, our first wave had gone ashore through a tremendous French bombardment. Our ships supported the troops as best they could, but there were few targets visible behind the dunes, so the French held even more than the usual defensive advantage. Nevertheless, our first wave had been spectacularly successful. The troops had landed just as rehearsed, then moved forward as if they were on parade, although enemy infantry and cavalry had been positioned to drive them back into the sea. Our comrades had fought magnificently and had forced the French from the beach in short order. On the way in, a few boats had been hit by round-shot and sunk, but most of the soldiers, heavily laden as we all were, had been rescued by boats designated for that purpose. It had been a well-organized operation. General Abercrombie seemed to know what he was doing.

Now I stood, feet braced and fingers flying as we surged ahead. Subconsciously, I adjusted the tempo of the tune to fit the oar strokes. Ahead, I watched the white lines of breakers crashing on the beach. They appeared wider with every passing moment. The merciless afternoon sun glared down on our armada of small boats as hundreds of sweaty sailors pulled us towards the beach. The 79th was part of the second wave, and contrary to our expectations, there had been no cannonade as we skimmed forward. To our right I spotted the site of my last battle, the disastrous Turkish landing. I could even pick out the spot where Mehemet Ali and I had thrown ourselves into the bay to escape the French. There on the very end of the peninsula was the familiar small fort, Aboukir Castle, now a key point in the French defences. With only a slight bump our boat slid onto the shore.

I stopped piping and waded into the shallow water. Ahead on the sand-dunes was our rallying point, a Camp Color flapping lightly in the breeze. I struck up the pipes again, and played "The Camerons' Gathering"[41] till all of our Company's boats had unloaded. Everything had gone like clock-work, due mainly to the exertions of the sailors making up the boat crews that rowed back and forth all day delivering the troops. We felt prepared for anything now. Each man carried, besides his weapon and ammunition, a large water-bottle and three days rations. Being the Company piper, I also carried the set of pipes loaned me by Captain Carnegie, and in lieu of a musket, a broadsword and Colonel Phélippeaux's pistol.

We formed up and advanced inland through the shifting sand past large numbers of dead redcoats -- many of them highlanders of the 42nd, easily picked out by their kilts and the feather bonnets which fluttered forlornly in the sea breeze. The casualties had been heavy, but eventually we came upon the French dead and it was reassuring to discover that they too had lost heavily.[42] The sight of several cannon taken from the French buoyed our spirits considerably, because only a defeated army leaves its guns behind. By the time we bivouacked, the entire British Expeditionary Force was bursting with pride. Finally, we had been led to a victory! A landing in the face of a well-prepared enemy is usually considered a guarantee of disaster. Our first wave had driven a large defending force from its positions and had pursued them inland even though, unlike the enemy, we had no cavalry. We had beaten the French Army, the most feared fighting force in the World! The next battle would be the 79th's chance to show what it could do.

～

The following few days we spent unloading stores and carrying out other frustrating work. The nights were wet and cold, and we had no tents. My experience bivouacking with *le 32eme Demi-brigade de Ligne* now became invaluable -- although I never mentioned how I came by my experience. French soldiers are amongst the most resourceful in the world. No Frenchman will willingly undergo personal discomfort while means to alleviate it are at hand. I had been one of Uncle Pierre's keenest students, and soon had our file of Camerons building "bivvies" out of palm fronds and other discarded rubbish. Captain Carnegie was pleased by my initiative and passed the word on to Colonel Cameron, and within an hour he had the entire regiment assembling bivvies, although they didn't rival those built by the Army of the Orient -- I assume because from birth British soldiers are used to hardship, and therefore have lower expectations of comfort.

Instead of advancing along the coast towards Alexandria, we seemed to be making the same mistake old Mustafa Pasha had made in lingering on a narrow neck of land while the French assembled to hurl us back into the sea. But for the storm, which had dispersed our fleet on the way from Marmorice Bay, we would have had horses for our cavalry and artillery. Fortunately, the Greek ships carrying our horses started to limp into port one at a time over the next week. In the meantime sailors had been brought ashore to act as laborers and drag the guns through the dunes. I thanked my lucky stars that my days as a 'Jack Tar' were over.

I must here explain the lay of the land because it was unique and totally dominated the conduct of any warfare near Alexandria. We had landed upon the western end of the Nile Delta, which is broken up by branches of the river. During much of the year most of the area is dry, or at worst, swampy, but when the Nile floods, the lower flats in the Delta become "lakes". Of course the floods cause shifting, so the lakes never retain their shapes. Behind the beach upon which we had landed a newly-formed lake was now discovered. We called it "Lake Aboukir". It prevented us from advancing inland and from being outflanked by the French. It also enabled boats to enter through a cut our engineers opened into Aboukir Bay thus supplying us from either flank. This speeded up our operation, but it still wasn't till March 12 that we finally moved westward towards Alexandria.

Among the three kilted Regiments in the expedition there was a fierce rivalry. The 42nd Black Watch were reasonably fine fellows in our eyes. After all, hadn't they been rebuilt using our men? The 92nd (who styled themselves "The Gordon Highlanders" because they had been raised by The Duke of Huntly, Lord Gordon's son) were a different matter. There was a bitter rivalry -- almost a hatred -- between the two units. For a start, one branch of Clan Cameron had actually joined the Gordons rather than the Camerons. This gesture during a time of intense competition for recruits had infuriated Colonel Cameron and our men. The Lochiel branch of the clan had joined the 92nd under the command of John Cameron of Fassiefern, rather than serve under our Colonel, Alan Cameron of Erracht, a kinsman with whom they had some family feud going.[43] When the 79th had been forced to recruit where it could -- notably in London -- the backers of the 92nd spitefully tried to have the 79th removed from the list of kilted regiments. Feelings between the two units had grown so strong that bar-room brawls had become common in England, and our regiment had eventually been exiled to the Isle of Wight -- far enough away to prevent more brawling.

Consequently our noses were badly put out on March 13, when the hated General Abercrombie gave the place of honor to the 90th "Perthshire Greybreeks" and the much despised 92nd. These two regiments were ordered to lead the advance during the Battle of Mandara. Both fought superbly and suffered heavy casualties driving the French out of their positions. For the rest of the day the enemy retreated slowly and skillfully, taking advantage of the rough ground and their great superiority in artillery and cavalry. Tangles of scrubby hillocks, palm groves, and swamps littered the otherwise flat area -- great terrain for defence. At last, around noon the 79th was ordered to pass through the 92nd, exhausted during the morning's fierce action. Catcalls and comments about "Cockney Jocks" greeted us as we advanced through them, but these only made us more determined. I know I piped my very best as we strutted through the hated 'Gordons'.

There were few momentous events during this, my first action with my regiment. In skirmishing order, we pressed on through the sandy landscape. Occasionally we fired volleys at barely visible French troops, then stood to reload under a hail of grapeshot from their artillery which was able to dart from place to place because of their horses. During most of this I stood beside, before, or behind my comrades rattling off reels and quicksteps to "invigorate the men" as Captain Carnegie commanded. Meanwhile we either advanced without artillery support or waited till the navy dragged up a couple of guns to blast away at the French who had already begun to retire.

During one of those pauses I heard a familiar laugh. I turned to discover Sir Sidney Smith cheering his men on as he helped sight one of the guns they had brought up. His black curly hair and hooked nose were the same as before, but he now sported a pair of enormous moustaches which curled like a ram's horns on each side of that unique nose. Smith saw one of our officers and gave him a low bow. "Congratulations on the gallant conduct of your men, Captain," he said. "Give my compliments to Colonel Cameron on their cool behavior under such heavy fire."

"Sir Sidney!" I called out, for I still was bumptious enough to have forgotten my lowly status.

Both officers looked about, our Captain about to tear a strip off the insubordinate 'Other Rank' who had dared to be familiar with an officer. Sir Sidney caught sight of me first and exclaimed with great joy, "MacHugh! What a great sight for sore eyes! How are you doing, lad?"

"As fine as can be expected under the circumstances, sir," I said waving at the musket balls droning by us.

"Captain, this piper of yours served with me aboard **H.M.S. Tigre** two years ago. A great help he was. MacHugh here has explored Egypt from Alexandria to Cairo and has communed with the Sphinx itself. I heartily recommend him to you as a trusted soldier and guide during your sojourn here in the land of the pharaohs."

The Captain thanked him for the advice then turned to me and said, "Come, Piper. Now it's time to earn your shilling." (He meant the bonus we pipers received when we enlisted.) As I turned to leave with him, Sir Sidney spoke quietly to me, "MacHugh, I haven't forgotten your services for me. Nor have I forgotten our dear Colonel Phélippeaux's promise to you."

Remembering my true status, I snapped to Attention and saluted. "Thank you, sir," I barked. Sir Sidney rewarded me with a warm smile as I struck up the pipes. That was the signal for the 79th to move off again. Memories and questions so filled my mind that I don't even remember what I played on that occasion.[44]

The French retreat had begun to look like a rout, and we were advancing rapidly when suddenly we topped a low ridge. Two miles ahead, our goal, the city of Alexandria, shimmered in the heat haze. In between lay a plain dominated by a long ridge upon which the French artillery now perched. Barely waiting to catch our breath, we pushed off towards an isolated barren mound of white sand, jokingly referred to as "The Green Hill". By now we had left Lake Aboukir behind. On our left was the bone-dry "Lake Mareotis". Our left flank was now open, and I must admit that I kept a wary eye upon it as we trudged on through swirls of salty dust. We were under heavy artillery fire most of the way, but it was strangely ineffectual. The roundshot either flew above our heads, or rolled along the ground in plain view. Soon we were receiving unusual orders such as "Open ranks right" and the ball would bound by, just missing our feet to pass on harmlessly through the whole regiment. We took and held 'The Green Hill' well in advance of our front line, but it was now obvious that any frontal assault would destroy our little army. Consequently, General Abercrombie ordered a withdrawal of the entire Army to the ridge we had captured earlier.

When night fell we felt pleased with ourselves, but disappointed that the 79th had not been given a more glorious role in the small victory. The 42nd had won immortal glory on the day of the landing, the 92nd had led the entire army today, while we had trudged along after the victory, getting shot at and eventually being ordered to fall back. I was young and foolish, and felt cheated by Fate and by Sir Ralph Abercrombie. Sixty of our men, I'll wager, did not feel that way. They were our casualties, eight of whom were dead.

That evening as we sat around our small fires on the reverse slope of the ridge I thought of my three friends of *le 32nd Demi-Brigade*. Had they faced us earlier today? Were they all well?

"Yank, I heard you palavering' with some Navy big-wig, the duke with the whacking great mustachios," commented Taddy 'Smith', breaking into my reverie.

"That windjammer is the famous Sir Sidney Smith," asserted 'Inky' Downton, another Londoner, a former chimney-sweep. "'ee's a bit of a rum duke, but the weevil-eaters all swear as 'ow 'ee's the real goods. Plugs alongside of 'em, 'ee does, and brave as the beast."[45]

"Is that so?" responded Taddy with a sneer. "**Sir** Smith, hmm? Must be my long-lost cousin," he laughed. "Say, 'ow is it that you gets to talk wiv him all chummy-like, Yank?"

I had long ago realized the futility of telling Taddy not to call me 'Yank'. It brought back memories I abhorred, but now I realize that in Taddy's eyes that name dragged me down to his level, so he persisted. "I worked for Sir Sidney when I was sent out here," I explained shortly. I had kept most of my adventures to myself -- for after all, who would have believed them? By relating the bald truth I could only have earned a worse nick-name -- such as 'Windy'. My mess-mates knew only that I had been sent as a piper to the Royal Navy, and believed that I had sailed the Mediterranean for the eighteen months I had been absent from the Regiment.

"What wiv you knowin' all about Egypt' and the Gyppos, Yank, and 'aving an in wiv the officers," commented Taddy thoughtfully as we bedded down that night, "and me 'avin' an eye fer commerce and all, we could do ourselves right proud in this 'ere place."

"I don't have 'an in' with the officers."

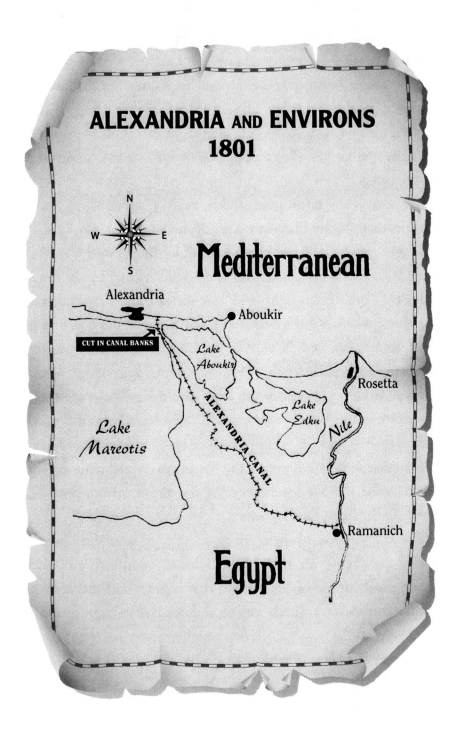

"Bollocks!" he snorted. "Captain Carnegie found you a bagpipe fast enough. Now this other Smith fellow comes along all chummy-like." I thought Taddy had fallen asleep, but in a moment he whispered, "Yank, did you know Inky's bin savin' up? 'Ee's a bit of a skinflint is our Inky. I can see great opportunities for you and me, mate."

~

Well before dawn on March 21 our outposts detected enemy activity to the left of the Regiment's position. We were stationed in the second line on the left flank of the Army together with the 50th Foot and the 27th Inniskilling Fusiliers. Colonel Cameron wheeled us to the left and we advanced to the banks of the dry canal bordering our Army's flank. After a moment of firing into the pitch black all fell silent. I heard Colonel Cameron telling Lord Hopetoun, the Adjutant-General, that it was just a feint by the enemy to draw our reserves away from the real point of attack somewhere else. He was correct, for a few minutes later firing broke out in the distance and grew to an awesome cannonade. The famous Battle of Alexandria had commenced! For the 79th it had also ended.

Alexandria was the battle that rehabilitated the British Army in the eyes of the world. General Menou, the French Commander-in-Chief, had elected to attack us in our positions along the ridge east of Alexandria. It became a savage infantry battle, although the French made great use of their cavalry and artillery. Nevertheless, our infantry won the day. On the right flank the 28th, 58th and especially the 42nd won immortal glory. Several other regiments also came into the action and earned themselves reputations as formidable fighting machines.

The 92nd "Gordons", marching off to besiege Aboukir Castle, managed to return in time to get in their licks as well. Meanwhile over on our left flank little happened even in the front line, while we regiments in the second line would have been just as effective making sand-angels or doing our laundry in the canal.

Some of our fellows went over to the 42nd's bivvies the day after the battle and returned shaken. "All our old 79th lads have been killed! Alan, Sandy, Neil -- they're all deid. There's only a handful of swaddies left, and they were all sittin' about in their bivvies greetin' their eyes out. The 42nd is all but wiped out!" It was true. The 'Black Watch' had been in action all day in at least three major scuff-ups, and had over 300 men killed and wounded. During the landing they had lost 200, so there could have been few left. It made us think. Much as we wanted to get a chance to distinguish the 79th, we didn't want to see all our friends killed. Of course, none of us considered for a moment that we might ourselves be killed.

News came back that several senior officers had been wounded. General Abercrombie had almost been sabered by a French cavalryman, but had been saved by one of the 42nd. Apparently he had also been wounded in the thigh, but it was not serious and he had been taken out to a ship to convalesce. General Moore, a Glasgow man whom all say was the hero of the day, was slightly wounded as well. The French lost two of their best Generals and about 4,000 men, so they took a pretty severe beating. All in all, Alexandria was one of the finest victories the British Army has ever won -- and at a point in history when it was vital to do so. We were all very proud of our performance, and even the 79th grudgingly admitted that Sir Ralph

Abercrombie had done a sterling job in rebuilding a fighting force that could defeat the scourge of Europe. We just wished that we had been given a larger role in the stunning success.

∼

It was during this period that sizeable numbers of Turkish troops arrived. A day or so before the Battle of Alexandria a rabble of several hundred "irregulars" were deposited on the beach. Recognized by everyone as totally useless -- except for pillage and rape -- this preposterous mob was banished to a spot well behind the lines where they could not get in the way. A few days after the battle several thousand more arrived. Amongst them were some real troops -- mostly Greek and Albanian. The whole force was under the command of 'The *Capitan Pasha*'. Seeing the Albanians, I kept an eye peeled for my old acquaintance, Colonel Mehemet Ali, but only caught sight of him from a distance.

Two days after the Battle of Alexandria I was called before Colonel Cameron. I had behaved well throughout my service with the 79th, so I was surprised and alarmed. The Colonel of a regiment is normally a distant and elevated figure whom Privates seldom, if ever, encounter. 'Old *Ciamar A Tha Thu*' was unique. [All highlanders called him this because of his habit of greeting everyone, regardless of rank, with the Gaelic salute, "*Ciamar A Tha Thu*?" ("How are you?")] Alan Cameron of Erracht was a tall, heavily-built man. His life story sounded almost mythical, thus he was held in awe by the men. After killing a fellow clansman in a duel over a lady, he had fled to America, becoming embroiled in the War of Independence on the Loyalist side. Cameron had first served as a secret agent and had been captured. No doubt, his size made him anything but inconspicuous. He had soon escaped and rejoined the Loyalists. Again he was captured and imprisoned, but made three more escape attempts, breaking both legs on one occasion.

Cameron of Erracht was one of the few men who raised a regiment and actually served with it. The 79th was more than a business venture with him; it was his passion. Detractors claimed he had chosen his English wife solely because of her wealthy father -- a source of funds to raise the "Cameron Highlanders". The fact is that Erracht had been a twenty-nine year old cripple (as a result of his escape attempts) when he wed the fourteen-year-old Miss Phillips, and his father-in-law had been so incensed at the marriage that he wouldn't let the couple set foot in his house for years. But Alan Cameron had eventually won over his reluctant father-in-law, as he won over most men he dealt with, and in 1793 the fortune earned by Phillips' Jamaican plantations was poured into raising the 79th. Our regiment was a close-knit family despite the odd mix of recruits drawn to it.[46] The Colonel's eldest son, Phillips, served as a Captain in the 79th, while six other officers and nine sergeants also bore the name "Cameron".

The redoubtable Lieutenant-Colonel Cameron met me outside his tent as he was about to set off on his rounds. In the background I caught sight of three burly sailors. "*Ciamar a tha thu*, Piper MacHugh? Ach! What sort of a lad are you? This morning I received **two** letters regarding you." He scowled down at me, as though sizing me up. "Where did we recruit you, lad?"

"London, sir," I replied with a gulp.

"Have you ever been in the Navy, MacHugh?"

"Not officially, sir."

"What do you mean, 'not officially'? I have here letters from two naval officers each demanding that I turn you over to them. Are you a good piper, lad?"

"Yes, sir. My father was taught by the MacCrimmons, sir."

"Damned if I'll give up a good piper -- even one who talks like a Yankee -- to please a couple of weevil-eaters. Captain Carnegie says you are a good player -- and well-behaved. He says he even supplied you with a set of pipes at his own expense. Claims you've never once been tipsy on parade. Must be some kind of paragon! Have you ever deserted, lad? Tell me the truth here and now, or by God I'll have you flogged."

"No, sir. I have never deserted, nor have I ever been a real sailor sir."

"No? Then how is it that Captain Creevey of **La Sangsue** wants to put you in chains for deserting to the French two years ago?"

"At that time I was working as a courier for Sir Sidney Smith, sir. I was officially a member of this Regiment, but he had taken me aboard **Tigre**, his flagship, and in the course of my duties I had been left ashore at the siege of Acre. I was captured by the French, but they thought I was -- well, balmy, sir. So they kept me till I escaped in Cairo and returned to Sir Sidney. Captain Creevey calls me a deserter because I was listed with **La Sangsue's** crew for rations. I believe he doesn't approve of Sir Sidney, sir, so this is one way of making him look bad."

"My God, what a tale! You don't mince words, do you, laddie? Well, I'd believe a piper and Sir Sidney Smith before some salty-arsed canvas-walloper any day. The second note I received was from Sir Sidney, himself. He claims you're so valuable that we'll fail to take Egypt if I don't send you to him immediately. So off you go with the weevil-eater on the left, the one sent by Smith -- and tell my friend I wish him all the best, but that I want my Yankee-talking Piper back as soon as he's finished with you."

Thus I arrived at Sir Sidney Smith's tent escorted by one of the sailors. "Ah, MacHugh! I can put you to much better use than hauling supplies," said Smith. "This evening I am to take a letter to General Menou on behalf of the Commander-in-Chief. I want you to carry my flag of truce because I need someone who can chat with the *grognards* we'll meet. With any sort of luck, we'll get into their camp, or even into the city itself, but whatever happens I'll feel most confident having someone along who understands not just the French language, but the French soldier. Therefore I put it a bit strongly when I begged Colonel Cameron to loan you to me for the day."

"Thank you for your confidence, sir, but I hope you won't mind if I bring up another matter while I'm here. Captain Creevey also sent Colonel Cameron a letter. He claims I deserted to the French and must be returned to him for court-martial."

"Oh yes. I am sorry about that matter. I wonder where he heard about your time with the 32nd Demi-Brigade?" he mused, twirling his mustachios. "By the by, your old comrades-in-arms came up against our lads the other morning. I am proud to say that the 28th Foot was more than a match for them. I hope no personal friends of yours were involved in that fracas because the 32nd Demi-Brigade lost heavily. Now about Captain Creevey's persistence! I must get that straightened out for you. The man does seem to have a very large bee in his bonnet about that incident."

That evening the two of us advanced from our lines, me carrying a white flag on a sergeant's halberd, Sir Sidney riding behind me with his usual affable smile, the very picture of grace, breeding, and extravagance. I had taken the afternoon to burnish my gear and brush my uniform. I must admit that I too cut a stylish figure -- my feather-bonnet at a jaunty angle, all its plumes dusted and fluffed out, my red coat and Cameron kilt almost spotless, belts and buckles pipe-clayed and burnished. I had even worked over my brogues and red and white hose, but that was a lost cause, because after a few minutes in the sand they were scoured and grubby again.

We approached the French outposts, and I made sure our white flag billowed in the breeze. Several scarecrows gathered around us, swaggering and full of bluster. It was easy to tell they lacked confidence, but thought they were pulling the wool over our eyes. I'd lived with French soldiers long enough to know all the signs. A young officer appeared. "Good evening to you," began Sir Sidney. "I bear a letter on behalf of Admiral Keith and Lieutenant-General Sir Ralph Abercrombie to be delivered in person to the French Commandant, General Menou," announced Sir Sidney, cheerfully in French.

The young Lieutenant replied that he would have to obtain permission from his commanding officer. In a few minutes a Colonel returned to say we would not be permitted to go any further, and that he was instructed to take the letter to the Commandant. As soon as he left, Sir Sidney dismounted and struck up his banter with the enlisted men who clustered around us. They were very curious to look me over. "What does he wear under his skirt?" one asked Sir Sidney with a snicker.

"Why don't you ask him that? He speaks French," was the reply.

The insolent fellow turned to me as though he was a great wit about to reveal the most original *bon-mot* in all history. (Every moron who ever asks me that question wears that same expression.) Gawking from side to side to make sure his audience realized just how witty he was, he swaggered up to me and said, "Scotsman, what do you wear under your skirt?"

I had a ready answer, but knew it would not lead to the good feelings Sir Sidney hoped to instill, so instead I laughed and pointed to my feet. "Shoes and stockings, of course." The others chuckled -- in satisfaction at the comedian's disappointment, I hoped.

One *grognard* pointed to the bare feet many of them sported, and said with a laugh, "You're doing better than us then, friend, for we have neither." I couldn't help but notice the difference in our appearances. I was clean and smartly turned out; they were ragged and dirty.

"No doubt you fellows will each be given a fine pair of boots when your supplies come up," said Sir Sidney.

"Supplies!" one scoffed. "The day we see supplies arrive, there'll be snow on the Sphinx!"

"And boots!" added another, "That's the day General 'Bondepart' will return to lead us to '***glory***'." At this they all hooted. The *grognard* felt encouraged to continue. "*Sacre bleu*! If we did get boots, we'd probably boil them to make soup. We've had no meat -- not even a decent meal -- since we left Cairo."

They were well under way now. "I feel sorry for you English -- and Scottish -- for you will have to live here when you take over this wretched land. It's a great pity so many of us -- both

French and Scottish -- will have to die before our Generals yield to the inevitable."

"There is really only one man who wishes to stay in this hell-hole -- our new Muslim, 'Abdallah' Menou,"[47] asserted one *grognard* bitterly. "We all long to return to France. But if we have many more scraps with you fellows, most of us will never again kiss a French woman."

"I've fought Italians, Austrians, Germans, Russians, and Turks, but by the Saints, I've never endured such a tussle as the other day against you Scotch buggers!" commented an old Corporal, nodding his head for emphasis. "I can tell you without fear of contradiction that none of us want to go up against you '*sans-culottes*' again."

I thought this was a good opportunity to score a point for us so I spoke up. "That was only one kilted regiment you faced. My regiment hasn't even seen action yet, but we look forward to our turn."

"You mean there are more of you '*sans-culottes*' waiting for us? *Sacre Bleu*! Your fellows in britches were tough enough for me. I'll stick to fighting Turks from here on."

The banter went on for some time in like manner. The common soldiers were poorly fed and clothed, they were all home-sick, and they didn't want to fight us again. Eventually the Colonel returned bearing a reply from the French Commandant, whom he claimed was General Friant. Disdainfully he pretended that General Menou, the French Commander-in-Chief, was still in Cairo and would be arriving with many more troops -- an obvious attempt to work on our credibility and morale.[48]

Sir Sidney and I said our "*Adieus*" and turned to make our way back. "The letter I delivered offered them repatriation to France in our own ships if they would agree to leave immediately," said Sir Sidney. "I judge by the tone of their messenger that General Menou has refused. Pity! You see, MacHugh, our army is stuck here on the beach unless we can trick them into moving. Our army is too small to capture Alexandria -- let alone take the whole of Egypt. Of course," he added, looking me in the eye, "you understand that is just **my** opinion, spoken in the strictest confidence."

∼

The days that followed were a mixture of tribulations and joys. At first, everyone expected the French to renew their assaults, so each night we slept shivering in blankets in our forward positions rather than behind them under canvas. Of course, the days were growing hotter while the nights remained bitterly cold. Strangely, I found that I suffered less than most of the Army -- possibly because of my upbringing on the Great Plains where the weather is more extreme. Each day was filled with hard labour -- bringing up supplies and ammunition, strengthening fortifications, and burying the thousands of dead. On the positive side, the Arabs had now concluded that we would be the eventual conquerors of Egypt, and began to supply us with all manner of food, setting up markets close behind our lines. This was nullified somewhat by the fact that we had not been paid for over six months, so our shopping lists were pretty humble. The other delight was the Mediterranean itself with its long sandy beaches. For some of us, every moment we could find away from our duties was spent naked in the refreshing waters splashing, swimming, and being rolled about by the surf. Even today I relish the memories of those few joyful hours.

On the 29th, eight days after the great battle, word was read out to each of the regiments that our Commander-in-Chief was dead. Sir Ralph Abercrombie, who had forged us into the magnificent fighting force we had become, had succumbed to his wound. The Army's morale turned to gloom. Most soldiers loved the man because of his concern for our welfare and our training. In the 79th it was now considered churlish even to mention Abercrombie's dealings with the regiment in times past.

The new Commander-in-Chief was Sir John Hely-Hutchinson, a surly, ungracious Irish politician, who had the reputation of being bright but lazy. Confidence in our leadership wavered. "Why are we still here?" many asked. "By now we should have taken Alexandria, and marched inland to seize Cairo," said the campfire strategists. I said little, but my previous experiences in two sieges suggested to me that our Army was ill-prepared to take Alexandria by storm. We had no siege artillery and too few men -- and before commencing a siege we must defeat the army that blocked our access to Alexandria.

The same day General Abercrombie's death became known I was again summoned by Sir Sidney Smith. "MacHugh, we have to make a second attempt to talk the French out of Alexandria. I'm taking you with me again. Keep your ears open, but say nothing of Sir Ralph's death. Everything we say and do must radiate confidence and optimism. I know you can play the part. Things are nowhere near as rosy as we will be painting them to the French," he warned. "Our Government has fallen. Prime

Minister Pitt is out, as is the Army's chief benefactor, the man who devised the whole expedition -- Lord Dundas. That dolt, Addington, is now Prime Minister, and Lord Hobart is the new Minister of the War Office."

I was astounded by his tone, and by his indiscretion in telling me such things. I was too young to know anything. of politics in those days, but felt that Privates should not be privy to such views. Even worse was to follow. "Poor Hely-Hutchinson! He's rather limited in military skills, but now he has Lord Keith and most of the other naval officers on his back for not getting things moving here. I have given him the benefit of my experience, but that hasn't convinced him to take action either. When he learned, two days ago, that the French fleet had left Toulon loaded with reinforcements to be landed here, it almost made the poor fellow quake. Lord Keith is demanding that our fleet be taken out of Aboukir Bay to intercept this force before it arrives The troops don't know it, but already several of our best ships have been dispatched from service here to patrol well out in the Mediterranean." Sir Sidney must have then realized he had been indiscreet for he added, "I only tell you this, MacHugh, so you will understand how serious our situation has become. I know you can keep your mouth shut." He finished by dipping his head to look me in the eye, an action worth more than all the warnings ever spoken.

When we reached the French lines I was both delighted and alarmed to recognize the shabby "uniforms" of the *32eme Demi-brigade de Ligne*! None of the soldiers we met were familiar to me, so I told myself I need not worry. Who would ever recognize me -- shaven, polished, and wearing a kilt? Sir Sidney presented the same demands as before: "I bear a letter for your

Commandant from our Commander-In Chief, the Capitan Pasha, and Admiral Keith, which I must deliver personally," but again we were forced to remain where we were while the letter was taken to the French Commandant.

"It shall be a grand event when we turn over this damned country to you fellows," said one young grenadier.

"Did you bring any brandy?" asked another. "I would escort you to that old bugger, Menou, myself for a bottle of good cognac." This was greeted by laughter.

"I have a couple of cases back in my tent," said Sir Sidney, a winning smile on his face. "If I had known you chaps had a taste for the stuff I would have brought over several bottles. You fellows will have to let me know what you need, and I'll look it up for you. Of course, if your officers will agree to our generous offer, you can simply go home to France and find your own cognac."

We all chuckled, but I felt uneasy. Something was wrong. I looked over the growing crowd. Several rows back a pair of the tiredest, palest blue eyes I have ever seen were fixed upon me. It was Uncle Pierre! His face was bleak. I stared at him and heard none of the banter that surrounded us. Then he moved towards me, pushing his way casually to the front. I have no idea what went on around us, but I know we gazed stone-faced at one another for a long time. Finally I came to my senses, and reached around to my knapsack to pull out some fruit I had been told to give to our hosts. Wordlessly, I passed them out to those nearest, saving the last one for Uncle Pierre. As I did so, I indicated that I had no more and stepped back a pace or two. He followed me. "Edmond?" he whispered, "It is you?"

"Yes, Uncle Pierre. It is."

There was a long silence as he bit absent-mindedly into the fruit. Finally he said softly, "I always wondered why the first words you spoke to us were not words I understood. You didn't **really** desert us after all then, did you?"

"No, Uncle, I was not really one of you. I did not want to leave the squad -- but --?" How could I explain to him? Another long silence.

"Jean will be pleased to hear of you. He is a Corporal now -- like me. You wouldn't recognize him, so big and hard, with a moustache like your officer there," and he nodded at Sir Sidney who was chatting away animatedly.

"What about Marcel? How is he?" I asked quietly.

"Marcel was killed in the last battle -- against you fellows -- over there," and he pointed to the redoubt behind us. I must have looked stricken, for he added, "Do not feel sorry, Edmond. Death comes to all soldiers." He paused. "Marcel did not disapprove of what you did. He only hoped you were happy."

Suddenly, the soldiers fell back and the French Colonel reappeared with a note for Sir Sidney. I have no recollection of what went on. I remember walking back with Sir Sidney who, mounted beside me, chatted on amiably about something while I saw before me only the face of my old comrade Marcel -- I didn't even know his last name -- an enemy who had saved my life, who had befriended me, with whom I had shared hardship and danger -- and laughter. I would never hear his griping again, nor share a bottle of wine with him. I'd always hoped that when the war was over --

～

A few days later it was rumored that an expedition would be sent to capture Rosetta and open the main channel of the Nile to our shipping. The 79th, one of the few regiments still at almost full strength, waited for the word. Then a terrible three-day sand storm, the Khamsin, lashed the Aboukir peninsula, tearing down tents and filling everything we owned with sand. When it ceased, two and a half battalions of our troops and 4,000 Turks marched off on the road to Rosetta. The 79th stayed where we were. I believe some of our officers were on the verge of tears, frustrated at again being ignored. I know I felt a pang as I watched the 2nd Queen's and the gallant 58th march out.

That afternoon while several of us were on the beach naked, playing like wee lads in the surf, Colonel Mehemet Ali rode by leading a column of his Albanians. He looked in our direction, for I believe he admired young men -- and was startled to recognize me. He waved his soldiers on and stopped for a moment. In the uniform of a Private, I would never have dared to start a conversation with a Colonel, but as a naked boy it seemed different somehow. "How are you, sir?" I asked. "It's good to see you again."

"So it *is* you? Are you an Englishman or a Mameluke today? It is very difficult to tell," he added with a hint of a smile, the only one I ever saw on Mehemet Ali's face.

"Actually, sir, I'm neither. I'm posing as a Scotsman -- but I'm really a Canadian, eh."

I realized he was staring at my Sun Dance scars. "Someday the Scotsman should come round to my camp," he replied. "I will be honored to have him share my hospitality. As you have no doubt heard, we are leaving for Rosetta. After we have captured Cairo possibly you and I will be able to visit." He smiled and nodded farewell then rode off to catch up with of his column.

"Hey, lads," came the voice of Taddy from the surf, "Yank knows a high mucky-muck in the Turks! Our lad **does** get around!"

"Bugger off, Taddy," I growled and returned to being a boy once again.

∽

On the evening of 13 April all those who were not actively manning the front lines gathered to see a great spectacle unfold. General Hutchinson had finally decided to flood the dry bed of Lake Mareotis. Between it and Lake Aboukir the high banks of a dry canal were all that prevented the two joining. I suppose the plan was to protect our left flank by flooding the low area -- as effective a barrier to the French as could be devised.[49] Unfortunately, the flooding would inundate many many acres of land reclaimed by the poor *fellaheen* of generations past. But it is ever thus, it is the poor who suffer the vagaries of war. We knew what was coming and gathered to watch. The engineers had dug four channels almost through the canal walls. These had been blocked by fascines. When the last fascines were removed the muddy water surged forward in waves which tumbled ten feet into the dry bed. Immediately cataracts formed, and within minutes three-hundred feet of embankments were swept away. By nightfall the flood had spread slowly westward to vanish beyond our horizon The next morning we were cheered by the peaceful sight of blue sky reflected off the waters on our left flank, replacing the ugliness of scrub brush and salt-flats. Even better, the 79th received orders to prepare to leave Aboukir to join in the advance up the Nile to capture Cairo.

∽

A SECRET OF THE SPHINX

On May 9th, in front of Ramanieh, a small fort on the Nile, we cheered when the French finally stood to face us. As it transpired, our army did much waiting, some manoeuvring, but little fighting. It was late afternoon when at last we advanced through the smoking wheat fields set alight by cannon-fire. French cavalry approached as we moved forward in line, so we "formed square" to repel a cavalry attack. No sooner had we done so than their cavalry swung to the left and right revealing a battery of horse-artillery loaded and ready to fire upon us. Squares are the worst formation of all to receive artillery-fire so we were immediately ordered to reform line.

As we did so, with our Light Company on the left flank as usual, I could see Mehemet Ali's Albanian cavalry advancing on their left. The Colonel was well out in front, mounted on his big charger. At that moment, the French horse-artillery fired a salvo, and down went his horse. Mehemet Ali scrambled to his feet, but the retiring French cavalry had seen his predicament, and several turned to the attack, galloping towards the lone officer. Our Company was moving into skirmishing order, so I dropped my pipes and sprinted towards him. Like all pipers, I was armed with a broadsword and a tiny *skean dubh*[50]. Unlike the others, I carried Colonel Phélippeaux's bell-mouthed pistol slung on my sword belt.

Mehemet Ali immediately shot down the first dragoon. The other three circled, brandishing their swords menacingly. I arrived just as one engaged him in a swordfight. Taking a deep breath to steady myself, I fired my pistol. The blast caught the nearest dragoon in the chest and knocked him off his mount. That turned the last dragoon's attentions towards me, and I stood back to back with Mehemet Ali fending off the Frenchman's mighty sabre strokes with my broadsword. The dragoon was both skillful and strong and would soon overpower me in a sword-fight. I must change the nature of the duel. I ducked under a blow, and as his horse turned, dropped my sword, and with both hands grabbed his belt and heaved. Down he crashed. I rolled him on his stomach and leapt upon his back, pulling the *skean dubh* from my stocking. Yanking his head back by the hair, I slit his throat. A Blackfoot war cry burst from my mouth as I leapt to my feet brandishing the small, bloody knife. Later I shuddered when I realized just how close I had come to scalping the poor devil!

Several Light Company men ran up. They gaped at the carnage. The corpse sprawled grotesquely on the crushed wheat, a huge pool of blood forming around the whiskered face. His horse pranced nervously before me. Seizing the bridle, I offered the reins to Mehemet Ali. Wordlessly, he took them and swung upon the mare's back. In the background I could hear our fellows firing a volley at the horse-artillery as it limbered-up and made off. Mehemet Ali recognized me with a start. "You?" he gasped. Then saluting perfunctorily with his sword he muttered, "Thank you, Ensign!" and rode off to rejoin his men. Armed as I was, and wearing an unfamiliar uniform, I must have looked like a junior officer. I stared after him in silence.

"Goddamit, Piper MacHugh!" came Corporal McAnch's voice, "Get back in line, you mad bugger!"

CHAPTER 14

"Reunions"
(Giza and Cairo, 28 June - 8 July, 1801)

My fingers danced upon the chanter as well as I have ever played. It was one of the proudest moments of my life, as we strode through the Giza Gate to accept the French surrender. The cocky wee tune was a march version of the old *piobaireachd* called "**The Pibroch of Donald Dhu**", and despite the summer heat, I had Captain Carnegie's pipes singing as I strode along behind Major MacLean at the head of one hundred Cameron Highlanders. The Egyptians were cheering, for they had come to the conclusion that we would drive the French out of Egypt, and they weren't unhappy with that. For after all, we British **paid** for our food and supplies.

The 79th had been chosen to be first to enter the enemy fortress that evening, and because of my knowledge of French, Colonel Cameron had chosen me to be the piper. We had cleaned up our worn and dusty uniforms for the parade, our first since we had arrived in Egypt, because we wanted to let the French know that defeating them hadn't taken the starch out of us.

Oh, it was grand to savor that victory! For the 79th it was a victory of endurance and marching rather than fighting, for our long campaign had included not one real battle. There had been only that trivial affair at Rahmanieh, a break in the long weeks of marching. The terrible heat, the thirst, and the agonies of dysentery had been our worst enemies, although the dreaded eye-infection called "ophthalmia" was my greatest fear. Some of our soldiers had become totally blind, and even more had partially lost their sight.

In those days, although I had engaged in several pitched battles in other theatres, I was too young to have any real understanding of strategy. However, like most youths, I thought I knew it all. Consequently, I concurred in the general feeling that our commander, General Hutchinson, was a bungling fool. We were all fire-eaters -- particularly those of us given no chance to distinguish ourselves to this point. I now blush to think how I went along with those 'experts' who castigated Hutchinson for his actions -- or as we considered it, his inactions. Later I understood that Hutchinson had used our tiny army of 4,000 brilliantly in forcing over 10,000 French (protected by Cairo's fortifications) to capitulate.[51] This he had accomplished speedily and with almost no casualties on our part. This bloodless success would enable us to rejoin the small force still investing Alexandria where we would finally put an end to the French adventure in Egypt. However, all we understood at the time was that we'd missed our chance to win "Glory" -- thus preserving our lives. I suppose this proves we were every bit as foolish as Bonaparte's *grognards* when it came to "Glory". Fortunately, our commanders were not as dedicated as Bonaparte in pursuing this phantasm.

Our detachment marched smartly through the Giza Gate, and the Major gave me a nod so I finished off the tune at the end of the part. He halted us and ordered a left turn. A French guard already formed up to our front presented arms smartly. We returned the compliment, and I was proud to see how well we performed these drill movements after months of neglect. Within moments we had fallen out and stationed men at the various points of importance. I was ordered to the battlements to pipe, in this way advertising far and near that the British were the new masters of the Nile. I will always remember pacing back and forth rattling off a series of regimental tunes as I looked out across the desert towards the Pyramids, golden in the rays of the setting sun. When I turned about I could see the minarets and towers of Cairo shimmering in the distance across the Nile and its islands. Sheik el Bekri's mansion stood there -- and Zenab, his beautiful daughter. If only she could see me now!

I had planned to finish off by playing "**The Camerons' Gathering**", but a fit of foolishness, or conceit, or both, came over me. For the last few days a melody had been forming in my mind. It came to me every time I thought of one of the most beautiful and endearing spots I had ever known -- The Sweet Pine Hills. Those three lonely peaks towering above the magnificent, endless prairie dominated the memories of my youth in that wonderful far-away land. So, regardless of protocol, that simple melody dominated by its own three "peaks" took control of my fingers as I turned to face the Pyramids glowing like a trio of triangular lanterns. Somehow while I piped this strange new melody, the three ancient symbols vanished to be replaced by those representing my youth. A moment I will never forget.

The two weeks following were full of activity for me, yet my chief memory is of frustration. Being a lowly Private, I had little chance to follow my own inclinations. Much as I wanted to go into Cairo proper -- what our people called "Grand Cairo" -- to find Zenab, it proved impossible. The French officers behaved very gallantly, visiting us at the Giza Gate, even bringing breakfast for our officers, and escorting them around Giza, old Cairo, and Grand Cairo. Because I could get by in both French and Arabic, I was kept busy as a translator for various expeditions, but that left no time for my own ventures.

One day, while we were in camp just outside of Giza, Corporal McAnsh arrived at our tent with orders for me and two other men to present ourselves, fully armed, before the Colonel. Inky Downton and Taddy were with me at the time, so they "were it". When we arrived outside the Colonel's tent I alone was called inside. Colonel Cameron loomed behind a tiny table. I saluted. "Piper MacHugh reporting, sirrr," I bellowed in the traditional style which had come back into favor since we had become parade-soldiers in Giza.

"*Ciamar a tha thu*, MacHugh. Stand easy. Here's an old friend of yours to see you. He has prevailed upon me to loan you to him for the next few days to do an important job."

From the shadows stepped a man in civilian dress -- James H. Flemming. "It's good to see you again, MacHugh. You really do materialize in a wide variety of *personae*. Colonel Cameron tells me this is your true self -- a Piper of the 79th."

"Which reminds me, MacHugh," interjected the Colonel, "what the hell was that tune you played the other night at our official 'take-over'? I was expecting '*The Camerons' Gathering*', but instead I got some weird thing I'd never heard before."

No one could get away with lying to Old *Ciamar A Tha Thu*, so I mumbled something about a melody just coming to mind at the time.

"Well, actually it wasn't bad. 'Pipie' even said he liked it. But next time stick to the Regimental stuff."

"Yes, sir," I replied standing rigidly at Attention.

"Relax, MacHugh!" bellowed the Colonel. "That's an order! At the moment we are three gentlemen -- an officer, an antiquarian, and a piper." As if to create this temporary equality he rose to stand with us, but his enormous size and presence emphasized the futility of the gesture. "Mister Flemming has called upon the 79th for assistance. I have agreed to turn three of my lads over to him till he finishes what he has been sent to do," boomed Cameron. "As well, I will personally do all in my power to assist in this momentous endeavour."

"We are to search for the 'Rosetta Stone', MacHugh," explained Flemming succinctly. "Sir Sidney Smith has advised me that the French have made secret plans to dispatch it to France where Bonaparte wishes to pose as the savior of historic treasures and a scholar of great consequence. Strangely enough, the so-called 'Rosetta Stone' has become the greatest prize of this war. Battles may yet be fought which will claim thousands of lives. The desert and disease will undoubtedly claim many, many more. Yet the **appearance** of Victory will be seen by the public to be possession of this scholastic jewel -- the key to ancient arts and sciences -- the Rosetta Stone. The world will soon be run by publicists and propagandists, gentlemen. Bonaparte has already discovered that fact, and he plans to use a chunk of basalt to prove he was the true victor in this Egyptian venture. But we have no intention of letting him have it.

"MacHugh, you are the only British subject who has ever laid eyes upon the Rosetta Stone. You are also an intelligent and resourceful fellow who speaks the languages of all the principals. Therefore you and I will be teaming up again -- thanks to Colonel Cameron -- to find the greatest treasure of the ages."

I was astounded by this development, but found my tongue to ask, "What makes you think we'll find it here, sir?"

"Sir Sidney sent a message from aboard **Tigre**. It seems the *savants* of Bonaparte's 'Institute' left Cairo about the same time as our Army left Aboukir. We tracked them down the Nile to Ramanieh, then by camel through the rising waters of Lake Mareotis and hence into Alexandria. There followed a comedy of errors featuring the unfortunate *savants*. General Menou, an absolute ignoramus with a distaste for learned men, wanted no extra mouths to feed, so he loaded them aboard **L'Oiseau** to make a run for France. Of course the vessel was immediately intercepted by the **Cynthia** and taken to Aboukir Bay. Admiral Keith, himself not the most gracious of gentlemen, sent **L'Oiseau** back to Alexandria, in this manner to increase the discomfort felt by Menou and his unwanted guests. General Menou, not to be outdone in churlishness, ordered the unfortunate scholars and their ship to set sail within fifteen minutes or be sunk -- *'foutre à fond'*-- was his elegant expression. Incredibly, the gallant Menou actually paced the battlements, watch in hand, as his guns were run out to send his own unfortunate ship and crew to the bottom. With less than five minutes to go, the *savants* -- realizing the depths of Menou's contempt for their intellect -- threw themselves at the capstan and assisted the *matelots* to raise anchor and get under way. Of course, ***l'Oiseau*** was once again captured. This time, Admiral Keith -- struggling for supremacy in the realm of callousness -- threatened to set the poor *savants* ashore on the barren coast before burning the vessel. General Menou, acknowledging at last his defeat in this duel of churls, ungraciously agreed to accept his shuttlecock scholars. Sir Sidney, always on the lookout for intelligence of benefit to the Crown, went aboard ***l'Oiseau*** to commiserate with the unfortunate fellows. Fortuitously, in the course of his conversation with Geoffroy Saint-Hilaire, the spokesman for the savants, let slip that the Rosetta Stone was still at the Institute in Cairo."

This last statement surprised me, for even my momentary conversation with Geoffroy Sainte-Hilaire had revealed a man very unlikely to "let slip" anything -- even to a smooth-talker like Smith. "As a result," continued Flemming, "Sir Sidney at once sent urgent instructions to me to appropriate this stone, the symbol of Victory in Egypt."

With a warm smile to Colonel Cameron, Flemming added, "Contrary to your kind words, Colonel, I am not an antiquarian, nor are any of my acquaintances here in the Levant. That is another reason why I was so desperate to secure the services of the one man amongst us who has actually touched the famous lump of basalt."

"Aye, Flemming, I'm entrusting you with three of my 'breechless boys'," replied the Colonel confidentially. "They're good lads, though possibly a mite too prone to wench and drink. You look after them, and they'll look after you." He turned to me. "You're the senior because you're the Piper. Take muskets and twenty rounds each with you. Sounds like you'll end up guarding treasure. Remember, MacHugh, this is no lark. You're on duty." So it was that Mister Flemming, Taddy 'Smith', Inky Downton, and I set off for 'Grand Cairo' one bright summer morning.

I was surprised to see how badly Cairo had fared in my two-year absence. Always dirty and noisy, even the poorest districts had nevertheless held a special charm. There had been something infectious about the city's vitality and street sounds, the intensity of the people and their aggressive approach to commerce. Then most of the buildings had appeared on the verge of collapse; now the inevitable seemed to have happened. We passed through whole areas now in ruins. "Rebels" had retaken the city and had been driven out after an intense bombardment by General Kléber a year ago. The plague had since swept through, bringing the population to a new low in numbers and morale. Our arrival had buoyed hopes amongst the survivors, but it was a sadly depleted Cairo I re-entered in July, 1801.

The Institute stood as before with French sentries outside. Mister Flemming was passed from officer to officer until someone was found who would authorize our entry. It was clear the French Army still controlled the city and were intent on reminding us of that fact. Once inside the gate, we found the beautiful garden unkempt. "Cor blimey!", was Inky's reaction to the dusty elegance of the great hall, and the high ceiling echoing his words. In search of the *major domo* alleged to be in charge of the building, Mister Flemming left the pimp, the chimney-sweep, and the Blackfoot warrior to gawk at the casual luxury of the place.

Taddy threw himself down upon a sofa to recline in comfort as he gawked around. "Great festering bollocks! Every other swaddie's dossin' down in the sand while we're livin' like old 'Farmer George' 'imself in a palace. Can you imagine settin' up shop 'ere, Inky?"

"Naw. There ain't no chimneys in 'Gyppo, Taddy. Besides, it's too damned 'ot."

"I reckon the rent fer this place is goin' to be low, now the frog-eaters are about to flit," continued Taddy. "What do you think, lads? Lets say 'up yours' to the Army and stay on 'ere."

Taddy's dreams were interrupted by the return of Mister Flemming followed by an ornately dressed *major domo*. When the latter emerged from the shadow, I was startled to recognize his face. At first I couldn't place him, and I was certain he hadn't noticed me. Who was this bland fellow? That impassive, baby face was somehow familiar. My mind went back to that night outside Zenab's door. Omar of Myra's *serradj*! The man carrying the sack that wriggled. He didn't recognize me though. But why should he?

"Eznik here stayed on to supervise the building after the Members left," said Mister Flemming. "He is willing to assist us in our task, although he claims to know nothing of the Rosetta Stone."

"The what?" asked Taddy, whose natural impudence was re-emerging under the command of a mere civilian.

"'The Rosetta Stone' is why we are here, lads. Near the end of 1799 the French newspapers were full of stories of this relic discovered near Rosetta. The story has since made its way into newspapers all over Europe, including those at home. Because of some inscriptions carved on the stone's face, many claim it to be the key to unlocking all the secrets of Egypt's ancient past. We have been ordered to find this stone and take it to England."

"What's it worth?" asked Taddy.

"If you are referring to cash-value, Private Smith -- very little. It's real value lies in its historic and linguistic worth. It will enable enlightened men to unlock those long-withheld secrets of the ancient civilizations."

Taddy and Inky exchanged bemused glances.

"It appears we will have to begin the search here," concluded Flemming, "so, MacHugh, I must ask you to tell these fellows what it is we are searching for. You see, lads, MacHugh is the only Englishman who has ever seen the Rosetta Stone."

My comrades looked impressed. I described the stone to them. When I finished I saw Eznik staring at me, his impassive face betraying nothing -- but his straying eyes revealed his secret. He knew who I was -- or who I had been.

We spent the next three days at the Institute searching first the building, then the grounds for the Rosetta Stone. It was a period of luxury for us after months spent in barracks, the holds of ships, and bivvies in the desert. To sleep each night on real beds in separate rooms was paradise, a glimpse into the world of the idle rich. Both the former pimp and the one-time chimney-sweep were enthralled by the opportunity, and took full advantage of it. Each night they entertained local harlots after dining on gargantuan meals supplied by Mister Flemming in his usual genial and generous style. Harlots were of no interest to me; I had my inner eye fixed upon a princess.

Every morning Eznik turned up, but would soon disappear somewhere about the place. I supposed that he spied surreptitiously upon us, possibly chuckling at our folly. I know that on the third and last day of our search, I had the old familiar feeling of being watched. I tried various ruses to discover who it was or where he was hiding, but to no avail. I wondered -- but no! Eznik had no way of knowing it was I who had slain his master and lover, Omar of Myra. The afternoon ended with no results to show for our three days of searching.

That last evening before we were to return to camp, I asked Mister Flemming for permission to seek out El-Bekri's house. He laughed and said, "Yes, MacHugh. I appreciate that you remained here on guard while your comrades whored and drank. Tonight's your turn. No doubt you have an old flame or two you would like to visit at the house of the Nahib al-Ashraf. But lad, remember you are just a Private now, and a lot has transpired since you were last in Cairo. Oh -- and MacHugh, if you see her, say 'thank you' to 'Jasmine' for me."

I found my way to the mansion, but it was dusk when I arrived. Everything about the house looked the same, although the neighborhood had suffered during the two years since I last slipped out the gate. I walked up to the familiar entrance, but found no one there. Although I burned with desire to see Zenab one more time, I realized it would be the height of folly to try to get in, so I turned from the house and sadly made my way back to the Institute. In doing so, I passed through the little park, now neglected and shabby. The hair on the back of my neck seemed to stand up, and at first I thought it was the memory of my deadly "big brother", Roustam Raza. Then I realized it was not a memory that haunted me; I was being shadowed by someone real. Spinning about, I caught a fleeting glance of a black figure disappearing among the trees. I unsheathed my dirk warily, but the sensation had already vanished along with the phantom. I returned to the Institute.

"What do you think of my plan, mate?" asked Taddy as I strode into the lounge. He was spread out luxuriously on one of the sofas while beside him knelt a harlot holding a bottle of wine. "I thinks as 'ow we should skip out of this soldier-business and set up our own School of Venus right 'ere. I'd be the nob 'cause I can find the doxies -- it's me perfession, you know," he added modestly. "Inky, you got some quid stashed away. You could run the place and keep it up to snuff. Yank, you're about the toughest little bugger I ever met. You could be the Flash Man and deal with the locals too 'cause you know the lingo." He waved his hand airily. "These rum digs will just go to wrack and ruin, once the Army leaves. If we don't snaffle it while the snafflin's good, then some damned froggie'll grab it. We won this bleedin' war, not them. It's only right that we stay on and profit 'stead of them."

"What'll we do when the Army finds us missin'?" asked Inky. "You've already bin kissed by the drummer's daughter, Taddy. That ain't a dodge' I want to try. We got some spankin' big drummers in the 79th![52] "

"Old fat-arse ain't goin' to catch me this time, Inky. No bloody oatmeal-snappin' colonel is goin' to flog Taddy Smith. If'n you ain't man enough to join us, Inky, that's your lookout," he warned. Turning to me he asked, "Yank, wot about you? Or do you fancy gettin' yer bollocks shot off fer King and Country and old hopper-arsed Cameron? You know yer way 'round 'ere. They'd never catch us in Egypt, and you know it. It'd be yer chance to repay me fer gettin' us out of that dicey business in London."

"You ain't told us about that one," said Inky, showing interest.

"No, well it wasn't much of anything," I broke in. "Just two young lads getting into a bit of a scrape." I hurriedly changed the subject. "Taddy, what makes you think you can take over the Institute?"

"It'll be easy, mates. Listen, the frogs are about to leave. Who's going to claim this place? That poxy little molly, Eznik? I'll slit 'is throat if 'ee tries anything. Just think, lads, three of us -- trained soldiers with our bundooks -- maybe aided and abetted by a few other free-thinkin' coves -- English or French, who cares? -- we can claim it and 'old it. Once we 'ave the place, nobody can get us out. As for the whores -- there's a million of 'em 'round 'ere, and who's better at selectin' whores than yours truly? In no time we'd have a top-flight knocking-shop and be in the quid! We'd be like Pashas -- real gentlemen!"

Fortunately, Mister Flemming returned at that moment, and the other two headed up to their living quarters. I went to the library which had become of great interest to me. If the French were going to leave all those books then I was going to claim one for myself -- I guess at heart I was just as larcenous as Taddy. Holding a candle in one hand, I was pouring over the stacks when I heard a sound behind me. I wheeled to confront Eznik. He wore his inscrutable expression. Smugly he said, "Caspar, you see I remember you." When I showed no reaction, a fleeting hint of disappointment flashed through his eyes. "You would like to return to Sheik El-Bekri's house." He said it as a statement. Still I did not answer. "I can arrange it, Caspar. Dressed as a Mameluke, and knowing a secret way into the palace, you could easily gain access and visit the lady of your desires."

"Why would you do this for me?" I asked. "You know I have no money. We haven't been paid in nearly a year."

"You misjudge me, Caspar," he said with an aggrieved expression. "Let me just say that I bear no love for the Sheik. It would not displease me if his womenfolk were to be defiled by an infidel. As I am not inclined to that form of pleasure myself, I would not be adverse to assisting a fellow 'Mameluke' to do so." Although he smiled smugly there had been no 'give-aways'. Eznik was telling the truth!

"When can you arrange it?"

"Right now. I have the necessary garments and weapons."

Eznik proved as good as his word. Within an hour we were standing outside the palace of Sheik El-Bekri behind a tangle of shrubs that had grown so high they obscured a narrow window covered by a grating. Eznik pulled upon the grate, and it swung upward without a sound. He smiled reassuringly and whispered, "It was always a good thing to be able to slip away from my master, Omar of Myra. I was delighted when he was killed, you know." His roving eye suggested he was lying, but to what extent I could not tell. Besides, it didn't matter, for my blood had begun to race with the thought of seeing Zenab.

I ignored all the instincts telling me I was a fool. I hadn't touched a woman since we left England -- and I had never touched a woman as beautiful as Zenab! I could almost taste her lips and her scent. Forgotten was the knowledge that she and I were not meant for each other. Once inside, I closed the grating and found myself in some sort of pantry. Obviously, the Sheik's kitchen-staff used this route to liberate some of his excess rations.

In the hall, remembering I had been transformed once again into a Mameluke, I swaggered as though I had every right to be there. I did encounter a few servants, but as is normal, they averted their heads and kept moving. In no time I was approaching Zenab's room. No one stood guard at the door. My heart leaped, and excitement overcame any remnant of caution. At the entrance I listened for a moment to insure she was not talking with someone. All was silent. I pushed the door open, and slipped in. Temporarily blinded by the sudden darkness, instinct made me step to one side of the door. I strained to see, and could hear the rustle of silks where the bed should be. "Who is it? Is that you, mother?" I recognized Zenab's voice.

I stepped forward. "It is I, Caspar."

She gasped, but before she could say anything I knelt beside her, kissing her lips. She responded instantly and warmly, at last stopping to gasp, "Can it really be you, Caspar? How can it be so?"

Feverishly, I removed my clothes and slid between the sheets with Zenab. It was as if I had never been away. We made love with such fervor that it was obvious that, like me, she had not done so for a long time. Neither of us could be satisfied, so the loving went on and on. When at last she fell asleep, the sun shone through the narrow openings high on the wall. Later, when her ladies came to dress her, I hid myself behind the curtains and hangings. Zenab yawned and explained to them that she had been unable to sleep, and had decided to stay abed till she felt better. Pouting, she imperiously ordered them not to return until summoned.

We spent the day slipping from love-making to talking to sleeping and back again. She had grown older -- more than the two years, but if anything she was more beautiful. However, there was now a fear in her eyes that even the excitement of my visit could not totally subdue. When I asked about it she laughed mirthlessly and changed the subject. "Caspar! You have no moustache!"

"In the British Army they are not the fashion," I explained.

"I have watched from the roof-top as new soldiers pass by in the streets. Mostly they wear red coats. I am told they are British. Do you too wear a red coat now, Caspar? You must have one with big golden buttons and much golden braid. A soldier as gallant as you must wear many such honours."

I couldn't bring myself to tell her that I was a soldier of the lowest rank, so I talked around the subject. "My uniform is different from most. I wear the kilt." Seeing her puzzled expression, I tried to explain. "It looks like a striped coat which hangs down to the knees, leaving the legs bare, except for our red and white stockings. On my head I wear a bonnet covered with black plumes. I think it's the finest uniform of them all."

"I have seen such men as you!" she exclaimed with delight. "They walk about with bare legs, and all the men and most of the women try to look under their strange short skirts. The French say they are men who have been cowardly in battle and have been forced by their dishonor to wear women's clothes.

But then the French also say these men in skirts are the fiercest of all the British soldiers. Obviously, one of these stories cannot be true, but now that I know you wear the 'keelt' I believe the second story with all my heart."

Later she commented wistfully, "You were right, Caspar, when you said Aboonaparte would not stay. He did abandon his soldiers -- and me -- and his French harlot. He cares for no one. You were right; Aboonaparte uses everyone. I hear stories that he has now become the Grand Vizier of his country."

Each time she dozed off, I had the opportunity to examine the mess I had gotten myself into. I had deserted. At least that's how the Army would look at it. I would be flogged, maybe even shot when I was found. Mister Flemming would find my uniform and equipment laid out neatly in my room at the Institute, and would believe the worst. The four of us were to have returned to the Regiment today. Perhaps I could stay behind with Zenab and resume my old life. Mamelukes were men of substance here, and ever since May when they had ridden into our camp while it still looked like we had real fighting to do, the Mamelukes had been the 'golden boys' of our generals. A Mameluke could probably carve a pretty good life for himself if he was careful. A lot of Frenchmen had made that decision, abandoning the bayonet for the scimitar. Certainly, I could hardly do worse for my own future than return to the British Army. I asked myself, 'What would my fathers do in my situation?' But I could not even imagine Donald Ban MacHugh or Sun Bull getting themselves into this mess.

Then Zenab would wake up and we would make love, and eventually I would go back to rationalizing the whole thing through again. Nevertheless, each time I concluded that if I followed the logical path I would betray the Regiment, and Colonel Cameron's trust, and Mister Flemming's word -- and myself -- and my Vision.

When night returned we had pretty well satiated ourselves. I announced my decision to return to the Army. I expected tears and recriminations, but she surprised me. "Caspar, my love, I know you must return to your Army. I now understand this is the way of men. But first I must show you something." Taking me by the hand, Zenab led me to a large chest. Proudly she opened it and there lay the green *yalek* and red *cahouk* and *charoual* she had worn on our outing to the Pyramids that day so long ago. She had secretly taken my gear and clothing from my room and stored it in this chest. In the bottom was the bag of gold coins I had hidden to pay for Roustam Raza's recruitment. She had not opened it, but I knew it must contain a tidy sum. "You were always disappearing and then returning to me," she explained, "Now I shall always keep something of you here with me, ready for your next return."

Donning one of my less flamboyant rigs, I hid the bag of gold inside my *yalek* and prepared to leave. At the door she took me in her arms and in an emotional voice pleaded, "You **must** come back to me again. I have planned a wonderful surprise for you, Caspar, my dearest. **Every** night I shall wait here for you because I **know** you will not fail me, Caspar."

Besotted by love, I agreed without hesitation. "I will not fail you, Zenab. I shall return no matter what perils lay in my path," I vowed. Yet even then I must have realized how impossible were my chances of returning to her after "deserting" the Regiment-- even if it was only for twenty-four hours?

I slipped into the hallway, and although there appeared to be no one around, before I had taken half a dozen steps a voice whispered behind me, "My lady wishes to speak with you." I wheeled and there was Jasmine's eunuch.

Wordlessly I followed, dreading what might come next. But the routine was the same as before, and within moments I slipped under the sheets with Jasmine. "Caspar, you rascal," she murmured, "you have pleasured my daughter for an entire day! I hope she is happy now. I know I would be." As before, our conversation was carried on in whispers, interspersed with mock sighs. It was one-sided, for each time I began to speak, she would simply thrust my face between her magnificent breasts. "I did not bring you here to pleasure me, nor I you, my sweet blue-eyed lion. Zenab is in danger. When the French leave, there will come a time of terrible vengeance. All who have supported them or are even accused of doing so will be sacrificed to alleviate the guilt felt by those who stood by and did nothing. The Sheik, Zenab's father, has much to answer for. During the uprising, when the rebels held the city, the people sought to kill El-Bekri. It may come when the British leave that we of his family will be required to pay for his subservience to Aboonaparte. Tell Mister Flemming that. Tell him also that his own life is in danger. There are those among the French who wish to even the score with Mister Flemming, and there are many hands willing to plunge the dagger into his back for favors and for gold. Tell him also that the stone he searches for is no longer in Cairo. Where it is I know not, but it was secretly moved from Cairo many days ago. Now go, my gentle lion." She gave me a long, almost motherly kiss on the lips, then whispered in my ear, "Allah bless you for the joy you have brought my darling Zenab." Then she pushed me out of the bed like a discarded pillow and rolled over. Totally befuddled, I got to my feet, donned my weapons, and slipped away.

I departed as I had arrived, and slipped across the street towards the small park on my way to the Institute. Mister Flemming and the lads would have returned to the Regiment by now, but I had to start somewhere. As I entered the familiar park I felt eyes upon me. In the center I halted, bathed in moonlight. Astride the path before me, blocking the entrance stood a dark figure. The silence was menacing, unnerving. At last the phantom stepped forward, but not a sound came from his booted feet. Shadow obscured his features, but the arrogant walk, the outline, both were familiar. The hair on the back of my neck stood up as the spectre approached, and I grasped the hilt of my scimitar.

At the edge of my patch of moonlight he halted. Slowly, deliberately, the spectre turned until the pale light illuminated his right side, revealing a terrible, twisted mask! A face shattered -- teeth and cheek-bone smashed into grotesque shapelessness -- the baleful eye sagging into the collapsed cheek! The spectre's thin lips twisted into a ghastly grimace of hate. Slowly the apparition turned its head the other way, revealing the handsome features of Omar of Myra.

"Good God!" I gasped softly.

"Long have I waited for this moment, Caspar." There was no doubt -- it was Omar's voice -- confident, threatening, but muffled and even more deadly than before. I warned myself to say and reveal nothing. He waited, then turned towards me, displaying the full horror of the once-handsome features, now broken into two aspects. "***You*** did this to me, Caspar. Your shot failed to kill me, but what it did was much much worse. Mutilated as I am, my master the Sheik El-Bekri, recoils in horror from me. His daughter, the despoiled Zenab, tried to stifle her shudders when I appeared before her, but I saw her loathing. My comrades shun me. Ever since that night at the Sphinx I have lived only during the hours of darkness. I have had ample time to contemplate my revenge -- on Zenab, on El Bekri, on many others. But above all, I will repay you, Caspar -- all in good time." With a melodramatic swirl of his black robes, Omar of Myra disappeared into the shadows.

~

"Damn it, where the hell have you been, MacHugh?" exclaimed Mister Flemming's voice when I slipped into the darkened Institute. I breathed a great sigh of relief. Maybe I wouldn't be listed as a deserter after all! When I failed to reply he actually chuckled. "I guess I know the answer to that question -- but tell me, have you seen the delightful Private Smith?"

"No, sir," I replied, and looking around, realized we were alone in the gloom.

"Private Smith has vanished. Deserted, I suppose. He didn't strike me as much of a bargain for the King -- hardly worth the shilling. I trust he received no more than a farthing -- still more'n he's worth. Tomorrow I'll return to Colonel Cameron only two of his men. Lucky for you that I trust you implicitly. I was certain you wouldn't desert, so I told Downton I had decided to scout around for another day. You should have seen his face light up! I hope he'll be fit to march tomorrow. Now let's hear what you've been up to -- and leave out none of the juicy bits!"

I related an expurgated version of my day of 'French leave'. "Jasmine gave me three messages to pass on. First, she says to watch yourself, sir. You have enemies amongst the French -- enemies who have sworn to kill you. Secondly, the 'Rosetta Stone' has been removed from the city sir -- some time ago, but she had no details."

"So Jasmine fears the French will hire someone to do me in, does she? I must keep my eyes open. As for the precious rock, we'll have to cast our net further afield -- although we are running out of time. I have been making discreet inquiries, and may have turned up a profitable lead. I believe our little *major-domo*, the esteemed Eznik, knows more than he lets on. His memory will improve at the sight of gold, I imagine," he added dryly.

"I will also pass on our lady-friend's news of the Rosetta Stone to General Hutchinson, but fear it will do little good. He sees me as a minion of Sir Sidney, and because he dislikes the Captain intensely, he will, I'm afraid, simply dismiss me and my report. Did you know that shortly after this expeditionary force left Aboukir, he ordered Sir Sidney into virtual exile from Egypt?" When I expressed my astonishment, Flemming went on, "Yes, as you may have noticed, Smith does have the unfortunate habit of wriggling into affairs that others feel do not concern him. He is very capable, but an incorrigible busy-body. It seems his continual meddling in minor military matters has put him on the outs with every senior lobster in Egypt. The final straw came when Sir Sidney told the *Capitan Pasha* -- the head Turk at Aboukir -- that only **he** had been authorized to offer terms of surrender to Menou at Alexandria. Despite Sir Sidney's claims that he was telling the truth, General Hutchinson ordered him to leave the beach forever and hie himself aboard ***Tigre***. Consequently, his messages to me now are all from off shore. There is no understanding that man! You were with Sir Sidney on both of his missions to the Commandant of Alexandria. Did he really offer surrender terms?"

I cast my mind back, but all I could recall was the tired blue eyes of Uncle Pierre as he forgave me and told me of Marcel's death. I had heard nothing of Sir Sidney Smith's discussion with the French staff officer. "I don't know," I answered lamely, "although I think I understand Sir Sidney. He believes himself to be one of the knights of old reincarnated, living by his own rules, pursuing his personal quest to bring justice to the world." I had been thinking about this for some time and felt very proud of my assessment.

"You said there were ***three*** messages," was Flemming's only comment.

"Ah! Yes, sir. The third is that Jasmine fears reprisals on Sheik El-Bekri's family when the French pull out." I wasn't sure how to approach this, personally the most important of her three messages. "She is particularly concerned about her daughter, Zenab, the one who was forced to become Bonaparte's mistress." I was surprised that I could utter the words so coolly. "There must be something we can do, sir, to prevent anything from happening to her."

"I doubt our government will want to meddle in the internal affairs of the Egyptians," he answered. "After all, the old Sheik brought any such retribution upon himself. You must have discovered, even at your early age, that everything has a price. It looks like the price for collaborating with Bonaparte will be collected with interest -- and soon. It would hardly be in our best interests to meddle in such affairs."

"But sir, Zenab is not to blame! She was forced to -- uh -- live with Bonaparte. She didn't ask to. We must do something to protect her, sir!"

"Ah -ha! You seem to be taking this personally, lad. Is there something you haven't told me?"

"Well -- uh -- she is a lovely lady, sir -- I mean -- as a person. She's innocent of any wrong-doing. It wouldn't be fair -- !"

"Yes, MacHugh, it undoubtedly would not be fair, but then what *is* fair? The war is still being fought, lad. She wouldn't be the first innocent to suffer from a war. But I will see what I can do. You understand that when the French complete their evacuation the Grand Vizier's horde will resume control -- for after all, it is their city. We will have no say in what happens here. Although the Turks have a few responsible troops -- the Albanians mainly -- the majority are just looters and thieves. There's also the Mameluke cavalry that joined us earlier. But the Grand Vizier doesn't trust the Mamelukes, and General Hely-Hutchinson will require them to help us escort the French back to the coast. No, I believe our best chance to help this poor girl is to request our friend, Colonel Mehemet Ali, to put some of his Albanians on guard outside the Sheik's palace. He is an independent customer, but he might agree to help. I shall also ask him to keep an eye on this place for us -- in case we have to return in our search for that damned stone." I must have had a worried look on my face for he added, "I'll do what I can, lad."

I sat gloomily trying to convince myself that he was right when I remembered one other thing. Putting my hand inside my yalek I drew out the small bag of gold. "This is the unused portion of the money you gave me to pay Roustam Raza, sir. I found it hidden where I had left it," I added, ashamed of my small lie.

Flemming looked at me in surprise. "You brought it back? I had given it up -- ages ago. This is only chicken-feed in our business, lad, but to a soldier it must seem a fortune. By God, you are an honest fellow! I shall return it and make mention of your conduct." Then briskly he changed the subject. "Now, from what you have told me, I have more immediate problems to deal with. Consequently I will not return with you in the morning as planned. I may have to ask Colonel Cameron to carry on the search in my stead.

"You do know the French will evacuate Cairo tomorrow, don't you, MacHugh?" There had been no French sentries loafing outside when I returned, but I had been too agitated about my own reception to let it sink in. "By this time tomorrow they will all be camped at Giza and on Rhoda Island," continued Flemming. "I wager we will all be leaving Cairo before the week is up. Sincerely, MacHugh, I truly hope you will have one last chance to say your farewells."

THE MacHUGH MEMOIRS ~ (1798 - 1801)

"Lt. Col. Allan Cameron of Erracht, Old Ciamar A Tha Thu"

CHAPTER 15
"The Sphinx's Warning"
(Giza, 13 July)

"Aha! there's the entrance to the tomb," exclaimed Colonel Cameron, consulting the sketch map given him by Mister Flemming earlier that morning. "According to Flemming, we should find the fabulous stone inside a sarcophagus in that very tomb." The Colonel had already shown me the note delivered by the hand of an Albanian soldier. I thought it considerate of him to do so, but as he had expressed it, "Dammit, lad, you're more involved in this bloody treasure hunt than anyone!" The note was as follows:

"My dear Cameron,

I fear I must presume upon your good nature and your well-known devotion to King and Country to beg you to embark upon a small expedition I should by all rights be carrying out myself. Unfortunately, I received a message from SSS requesting -- nay, demanding -- my presence, and have made arrangements to leave ere you receive this missive.

It might be best if you employ the two fellows I returned to you the other day, as they are both sound chaps -- unlike their erstwhile comrade, 'Smith'. The lads are acquainted with the major-domo who provided the information you will find enclosed in the form of a map, hand-drawn by yours truly. The lump of basalt appears to have been secreted by its late owners in a sarcophagus where X marks the spot. If, my dear friend, you could find it in your heart, and have sufficient time upon your hands, I humbly beg that you carry on in my place.

I advise caution as some miscreant made an attempt upon my life last night. Someone entered my chamber at the Institute by stealth during the early hours, and deposited a rather large and distinctly unfriendly cobra in a sack in my bed. Fortunately, I had been unavoidably detained elsewhere at the time.

As ever,
Antiquarian"

As a result of this message, armed with a sledge-hammer, I trudged behind our Lieutenant-Colonel. Alan Cameron was a uniquely personable man. He had few, if any, pretensions, and although he was a Colonel, seemed to care only marginally for the trappings of rank. 'Old *Ciamar A Tha Thu*' was more like a father to the 79th than its commanding officer. Blunt and forthright himself, he expected others to be the same. Consequently, he chatted away with me as he would with any of the officers. He was an easy target of ridicule because of his huge body, his booming voice, and his bluntness. Several men of the 92nd who had accompanied us out to visit the tombs snickered behind their hands at his gigantic girth and his easy manner. Few of them, I reflected, would have liked to stand up to him in his youth. When he got himself stuck in the narrow entrance of one of the tombs a shower of ribald advice and observations descended upon us, but by the time I freed him we were alone.⁵³ "Goddamned Gordons!" he growled, "laugh at a man's arse, but afraid of his face! Comes of servin' with Fassiefern." He would never get over the hurt of seeing other members of Clan Cameron serving with the 92nd.

When we arrived at the entrance the Colonel lit a torch and led the way down the narrow defile until we were crouched at the bottom staring at a sarcophagus. It was made of heavy stone blocks, and was large, although not large enough to hold the Rosetta Stone, in my opinion. But then who was I to say so? Our torch revealed hand prints on the dusty top. "Well, lad, now's our chance to discover if this bloody sarcophagus contains your famous rock. If Flemming is right we'll call up Downton with the camel, load it up then and bugger off." He heaved at the cover, but it did not yield. "Someone has already been here and pried this thing open," he commented as he studied the edge which was ever so slightly ajar. "They must have used this." He pointed to a Mameluke spear on the floor behind the sarcophagus. "Bloody tip's been snapped off. Looks like we brought the wrong implement for the job, MacHugh. Well it can't be helped. Take a whack at it, lad." His voice echoed in the cool chamber.

"Yes, sir. But -- uh -- this is centuries old. I mean -- uh -- to destroy this -- seems so --"

"Aye, but think what we're after, lad. King George's representative himself has ordered me to find that damned rock which is even more valuable than this sarcophagus. I have no choice -- and neither do you, laddie. Swing away and Devil be damned!"

I raised the sledge and brought it down with a mighty crash. The sound echoed in the narrow confines. Again I tried. The sole result was the reverberations off the walls. In the distance I heard voices. "Here, give me me the damned hammer," the

Colonel ordered, "I should have brought big McGraw instead of a wee piper." He hefted the sledge and brought it down with all the strength of his enormous body. The crash was louder than ever, and amid a spray of splinters, the top of the sarcophagus shattered and fell inward. He handed me the hammer and bent to look inside. The flickering torch revealed nothing, so he pulled off a chunk of the broken lid and leaned forward to peer again. A sudden flicker of reflected light darted at his face from the depths of the sarcophagus -- a cobra!

"Goddam!" he exclaimed and flew back faster than I could have expected. "Goddamned serpent just missed me!" He seized his broadsword and I took the torch in hand. Swaying slowly to and fro was a cobra. Something was wrong with it. Then I understood -- a portion of the shattered lid had pinned the snake by the tail. It's deadly strike had been shortened. The sword flashed and the cobra's head was lopped off. Hideously the body thrashed about then collapsed. I shivered, but the Colonel looked absolutely calm again. I believe the same thought struck us both. "Laddie, someone tried to kill me -- or Jamie Flemming, for it was he who was supposed to open this."

He was right. The cobra could not have gotten into the sarcophagus by itself for the lid had been very tight. Inside, a sack was visible in the light. Someone had pried open the lid, dropped the deadly sack inside, and lowered the lid again before the cobra had escaped. The next person to pry open the lid would be met by an angry and deadly snake. The spear must have been left for the intended victim. Needless to say, the Rosetta Stone was not inside the shattered sarcophagus.

"Well, at least I'll take back a souvenir," commented the Colonel hefting a chunk of the lid and examining the inscription.

"I say!" exclaimed a scandalized voice from above. "Did you chaps break into that sarcophagus?" We looked up to see several faces, white in the torchlight, peering down at us in dismay. "That sarcophagus is thousands of years old! It has withstood the ages, sir!"

"Aye, I'm just appropriating a wee souvenir," '*Ciamar a tha thu*' replied jovially.

Another voice, barely holding back a laugh, exclaimed, "Ah! Lieutenant-Colonel Cameron, it's you! How good to see you, sir. Thought it was a bunch of grave-robbers, sir -- came down to investigate. Cheerio, Colonel!" They waved and disappeared with their torches. I wanted to follow them. This was not a place for me. I made to climb up when I heard the same cultured voice in the distance exclaim, "The man is a veritable barbarian -- a Goth, I tell you, a bloody Goth!"[54]

~

Our return to camp with two chunks of the desecrated sarcophagus[55] was unusually silent so it was inevitable that my thoughts returned to Zenab. While I had been far from her charms common sense had prevailed, and I realized she did not love me as I loved her. I even recognized that we shared no basis for a lasting relationship. Yet ever since my return to Cairo, she had constantly been on my mind, and these rational conclusions no longer mattered. I specially recalled her parting words, "I have planned a wonderful surprise for you, Caspar, my dearest".

Today's shared adventure had created an unusual bond between Colonel Cameron and me. Although I did not want to use this bond to gain undue advantage for myself, I was desperate to see Zenab again. After all, I had promised her I would return. So I summoned up the courage to make a request of *Ciamar A Tha Thu*. "Sir, could you give me permission to go into Cairo for a day? I have a friend there, and it may be our last chance to -- uh -- get together, sir."

"Aye, MacHugh, you can have tomorrow to go and see your girl. It's damned easy to do something rash over a woman. I can vouch for that." After a moment he added, "Damn it, lad, you'd better leave this afternoon instead -- soon as we return to camp. Mind, MacHugh, I don't want you telling the lads I've gone soft -- and make sure you get back on time. We're marching out first thing the morning after to escort the Froggies back to the coast." At camp he scowled and added, "You'd better go armed. There's no telling what's happening in the city now with the Turks in charge. Tell Sergeant Cameron that I've sent you to bring back that goddamned Private Smith. I'll give the little bastard one last chance before I list him as a deserter."

So while the other swaddies were scoffing their rations I was hurrying back to Cairo. It seemed a long hike -- first into Giza, then across the bridge to Rhoda Island now crowded with French troops, there to find a boatman willing to ferry me across to the north bank for a few coins, and finally the trudge into 'Grand Cairo'. I could not help but wonder what Zenab's parting plea had meant. "I shall wait here for you every night. I have planned a wonderful surprise for you."

When I arrived at the Institute it was dusk, and the place seemed deserted. No one stood on guard outside, so I went straight in. At Mister Flemming's suggestion I had left my Mameluke togs in the room I had been using. As I climbed the staircase I sensed I was being watched. On reaching the landing

I looked around but saw no one, so continued on to enter my room. I pulled the dresser away from the wall and withdrew my clothes from my hiding-place.

"So you've decided to join up wiv me, 'ave you, Yank?" Taddy stood in the doorway. He carried his musket and was still in uniform, but looked scruffier than ever. He had been drinking.

"No Taddy. I'm here on my own business," I replied as I donned my Mameluke clothes.

"This 'ere is goin' to be a rum dodge, Yank. 'Oldin' on to this place'll be a bit dicey at first. Reckon I'll have to skewer a few Gyppos 'fore they learn to leave us be, but 'old on I will. I can go it alone, you know, Yank, but I'd sure as 'ell like you and me to stay mates."

"Colonel Cameron told me to bring you back when I return tomorrow, and all will be forgotten. You're a lucky devil, you know. Any other C.O. would have had your guts for garters," I said as I changed into my Mameluke rig.

"I ain't goin' back, Yank. All these years I've wanted to be somethin', and I'll never get a chance like this again -- my own knockin' shop. For once in my life I'm going to 'ave the bollocks to take what I want and 'ang on to it."

I suppose I should have told him just how preposterous his dream was, but all I said was, "Well good luck, Taddy. I'm off," stowed my uniform and musket, and left in pursuit of my own preposterous dream.

~

I used the same method as before to enter El-Bekri's house, and within the hour had slipped into Zenab's chamber. When my eyes became accustomed to the dark I realized that her bed was empty!

Alarmed, I tip-toed towards the back of her apartment. Suddenly a form in Mameluke garb appeared before me. I reached for my scimitar as the figure stepped into a shaft of moonlight. "Surprise, Caspar!" It was Zenab! She was wearing my green and red outfit. While I gaped in silent astonishment, she explained brightly. "Surely you remember! I promised you a wonderful surprise upon your return. We shall do as we did once before. Tonight I shall ride out as a man -- with you. I have arranged to have horses waiting for us, but we must leave quickly. It is a long ride."

"Where are we going?" I asked.

"To visit my sister, the Sphinx."[56]

Looking back, I am amazed that I didn't put my foot down and prevent the excursion, for I had entered the house with carnal thoughts in mind. Why I acquiesced so easily to such a foolhardy scheme I'll never know. I suppose it really was love -- that and my delight in seeing her take charge of her own destiny. The result was that within a couple of hours we were approaching the Pyramids, leaving the Nile and its cluster of squalid towns behind us.

The moon gave everything a magical luster -- sand, sky, Pyramids, and the Sphinx itself. The face gazed at me, as though sorrowing. An owl hooted once, but I was too besotted to pay it heed. I was in love with the sight, and the mood, and Zenab. As we approached, I realized she had been riding quite well. She must have practiced somehow. Now she spoke, and I still remember every word. "I know the great Sphinx has a special meaning for you, Caspar, and that it knows your many secrets -- ones that will never be revealed to me. I have come to believe she also holds a secret for me. Something of great importance to me will occur only after I have visited the stone cat with a woman's face. My dreams have not told me what will happen, just that it is fated -- but first I must touch her."

We halted and gazed up into the sad, enigmatic stone face with its backdrop of hundreds of stars. "Caspar, you must make love to me on the creature's back," she whispered. Obediently I led the way around to the rear of the colossus and assisted her to climb to its back. Immediately she lay down and beckoned to me. We made love tenderly, with the moon and the stars as witnesses. Our union was long and gentle, and when I finally rolled onto my side I heard in the distance the urgent cry of an owl.

"That was wonderful, Caspar -- but it was not all I expected." Seeing my hurt expression, she hastened to assure me. "You were wonderful -- but I expected something else to happen. My premonitions have told me to expect the most important event in my life to take place while we are here. But the Sphinx still has not revealed her secret to me."

"Sometimes premonitions and dreams are false," I lied to calm her. After a short silence the owl hooted once more, urging me to continue. "We cannot live our lives trusting only in such dreams -- or in Kismet." It was as though the Sphinx was speaking through me, finally revealing a secret it had withheld for so long. "Visions can only **suggest** the path to take. It is by our actions that we chose the path we follow. Kismet tells only what might be, not what must be." I could hardly believe this was my voice.

"Kismet tells me I must make the most important decision of my life -- now, while I am here," she whispered with a thrill in her voice. She had not been listening to me. Zenab had heard her own secret from the Sphinx. "Now I understand the secret! I must no longer be a meek pawn of others. I must make my own Kismet! Very well, great one, I will do so! In fact, I have already done so!" She turned to me. "I have decided that you shall be my husband, Caspar. You must take me to the British camp and there wed me in the tradition of your people."

The Owl cried urgently -- this time a warning!

"You must marry me now, Caspar," she said earnestly. "The British are all-powerful -- even more so than Aboonaparte. The Imams will not dare to follow me into your camp. They will not dare to interfere with the wife of one of the British leaders -- especially when I have wed you in the British manner."

"But you don't understand, Zenab. I am only a Piper! I'm not an officer. The Army will not allow me to marry, let alone take you away with me."

"But I thought you loved me, Caspar! I can grow to love you and be a dutiful wife. My mother has warned me that the Imams plan to punish *me* for what Aboonaparte did to me. The French have already left, Caspar. The British will soon leave. You must not abandon me! Take me to the British camp. If you will not wed me then I shall wed one of the other Pashas there."

"It is too late for that, Zenab," said a muffled voice.

I rolled over in time to see a dark figure above me and feel the sword tip at my throat. Suddenly, the back of the Sphinx swarmed with dark forms. The spectre holding the sword at my throat turned his face allowing the moon to illuminate it. The hideous misshapen mask of Omar of Myra twisted into a grimace of a smile. "At last! I have waited two years for this moment," he mumbled.

As someone struck me on the forehead and I lost consciousness I heard the plaintive cry of the Great Snowy Owl.

CHAPTER 16
"The Night of Many Reckonings"
(Cairo, 17 July)

"Today is the day you will remember for the rest of your life, Caspar. This is your day of reckoning," announced Eznik. His face was expressionless so I knew he spoke the truth. "Master has instructed me to take you to watch his triumph. You will not enjoy it as much as I, but then who enjoys their own degradation and agony?"

I had expected to be murdered, or at least tortured, but apart from a couple of beatings by a pair of identical thugs nothing else had been done to me. Like Marcel and Jean, what I feared most was to be sodomized. In vain I told myself there could be worse things. For the last three days, unfed and chained in a dungeon, I had tried to display a bravado I did not feel. Two days ago distant trumpet calls and a hint of the pipes had announced our Army's departure. Since then all I had heard from the streets was a babble of terrified voices punctuated by occasional gunshots, and shrieks of fear and pain. Had the Turks been let loose on the city? They had been recruited with that as their promised reward.

Where was Zenab? I had not seen her since our capture. Had Omar of Myra hurt her? He had loved her once -- if ever he had been capable of love. Zenab's premonition had been right, "the most important event of her life" had taken place with the Sphinx, but not as she had interpreted it. She had tried, too late, to seize control of her own fate. Now I understood the fear in her eyes. Her demand that I marry her had been a last desperate lunge to avoid the destiny she had foreseen. The gentle Zenab had fought a better fight than I had.

Why had I not recognized the repeated warnings of the Great Snowy Owl? Had love deafened me? The owl, my very own Spirit Guide, had come from the other side of the world to warn me, and I had ignored him! My wise Spirit-Brother would probably never guide me again, for he must feel shamed by my foolishness. My new friend, the Sphinx, had advised me to make my own decisions when he spoke the words, "Kismet tells only what might be, not what must be". Zenab, sister to the Sphinx, had understood that truth, while I protested that it was impossible. How blind I had been! Mister Flemming too had warned me. Friends and allies from every corner had tried in vain to help me, yet I had ignored them all because of my foolish and hopeless love for a woman who did not love me. What an ass I was!

My mental agonizing was interrupted by Eznik. "Master once told me he would rather be dead than mutilated. Your shot smashed his beautiful face, distorted his voice, and crushed his pride. Kismet would not let the musket ball veer so much as a hair's breadth to relieve him of his life. Such is Master's Kismet. It worked its evil through you, Caspar, but tonight it shall return evil for evil." Eznik's philosophical mien vanished and he almost sobbed, "You devil!" and hit me as savagely in the face as he could. Eznik relished the pleasure of it, and slapped me till he was exhausted. He drew his knife and looked longingly at it. "Master has threatened me with death if I should harm you in any way, but it would almost be worth his wrath to cause you to writhe and shriek with agony. However, that time will come, for he has vowed that you shall be made to suffer long and bitterly before you die. But first your degradation must be complete." He punched me as hard as he could in the stomach, but he was tired and I was prepared, so it was hardly worth his while. "Evil devil!" he gasped, "if only Master had let me come with him to kill you that night at the Sphinx, our lives would have been so different. But it was fated. Kismet will reward Master and me, and will punish unbelievers. It has already begun to punish that whore, Zenab. She has long fascinated Master, and I was forced to watch as she ensnared him. For years she held him under her spell. Now for the last three days he has been entreating her to reward his mistaken love in return for her life, but in her vanity she has refused him that which he could have taken by force. At last her spell has been broken! Within the hour Zenab, the wicked temptress, shall pay the price for her sins."

Eznik regained his impassive confidence and signaled the twin thugs to unchain me. In my weakened state I put up no fight, for I knew the little bastard might lose his self-control and use that dagger on me. In a moment they retied me so I couldn't move a muscle, and dragged me out of the dungeon and into a cart. It was evening. I rode through the streets with agonizing slowness, the object of curiosity. "He is a Frenchman," announced Eznik, so the passers-by threw rubbish at me and hurled insults. Eventually I began to recognize the streets, and we arrived in front of El Bekri's palace. I was dumped from the cart and dragged to a pillar in the courtyard. There I was tied, after being stood on a stool which raised me above the crowd -- for what reason I did not then understand.

The crowd swelled until it filled the courtyard. Surprisingly, they showed only marginal interest in me, and although at first some threw stray rocks and bits of garbage at me, Eznik's twin guards moved menacingly in front to protect me. Within moments a group of Imams wearing ceremonial robes filed into the square. Of all the hundreds of devoutly religious leaders these had been selected for their lack of humanity. Each cruel face proudly displayed the hatred and fanaticism for which he had been chosen. I have never before or since seen such an array of savages posing as the representatives of the love of the Almighty.

The Imams halted their procession of hatred, and their leader called for the owner of the palace. With surprising humility El Bekri appeared. But the mob wanted more and began chanting, "Bring forth the whore! Bring forth the whore!" The leader and nastiest of the Imams called for order and waited for the crowd to lapse into a reluctant silence. Stepping forward to claim total control of the ritual savagery, he demanded to speak to "the mother of the harlot who has defiled herself with the French infidels". A woman in black stepped forward. It was Jasmine -- at least I assumed it was she, for the figure, although shrouded in the traditional manner, seemed somehow familiar. Again the mob screamed in frustration, "Bring forth the young whore! Slay the daughter!"

Obviously relishing his role in this obscenity of 'justice', the Imam waited until the mob had once more lapsed into obedient silence before confidently demanding, "Bring forth the vile debauched woman!" This time the mob burst into frenzied screeches of approval that went on for minutes.

Far back in the darkness a pathway through the mob began to open as if by magic to reveal my darling Zenab. Defiantly shrouded in white, she stepped forward with a dignity so noble that the mob was soon shocked into silence. I could stand the injustice no longer and shouted, "Zenab! Zenab!" I was about to shriek "I love you!" when she looked my way. I swear I saw her face through her white veil! The serenity of her smile stunned me. She raised a finger to her lips like a mother hushing a restless child. Obediently, I fell silent. She lowered her hand and the vision of her face faded. Again I saw only the shrouded figure, but I knew she had fixed her eyes upon me. Even as Eznik forced a rag into my mouth and tied it behind the pillar to keep my head upright, I could see her steel herself as she turned to face the furious but confused horde.

The Imams took turns accusing her of being a whore, of willingly consorting with the French infidels, of bringing degradation to "all men of the Faith" and many more such lies. These "holy" men shook their fingers at her, and lied again and again. They spewed their accusations at the defenseless girl, the victim, not the perpetrator of their imagined crime. "Do you repent the multitude of sins you have committed?" one finally demanded.

"Yes, I repent my sins," she answered softly.

They turned to Zenab's father, the despicable Sheik El-Bekri. "What do you have to say in defense of your daughter?"

"I disavow my daughter's conduct! Until now I remained unaware of her degradation. Had I known, I would have taken steps to end it. This woman has brought indelible shame upon me! She is no longer my daughter! I disavow her completely."

I gasped and struggled to no avail. Silent and totally helpless, I watched the Imams tear off Zenab's head-dress to the accompaniment of the screaming and snarling of the mob, once again captivated by the ritualistic savagery sanctified by these liars and murderers. Two Imams seized her hands and stretched her arms outwards while she was pushed to her knees by a third. A greasy brute wielding am enormous scimitar stepped forward, surveyed the mob to demand his moment of importance, then flexed his massive arms as he drew the scimitar above his head. For a brief moment he focused his attention on the angelic form beneath him. I tried to close my eyes, but my heart would not permit it. He tensed his mighty arms and the shining scimitar flashed downward. My gentle Zenab's head rolled onto the stones in a rush of blood.

The crowd, cheered and chanted like victors at a football match. I am not sure what transpired in the moments after Zenab's murder, and only vaguely recall the mob dispersing to follow the 'Holy Men', chanting all the while, until only I remained, sobbing silently in front of El-Bekri's palace.[57]

～

"Zenab has now paid for her rejection of me. It was I who delivered her into the hands of the Imams," came the muffled voice of Omar of Myra. He stepped out of the shadows beside me. "El-Bekri has just begun his torment. His debt to me is so extensive it will take the remainder of his miserable life to repay. However, now is the time to settle my account with you, Caspar."

The thuggish twins took me down from my perch and threw me into the cart. After a short journey it rumbled to a halt, and I was dragged out. Although it was now pitch-black, I recognized the Institute. My feet were still tied, so they dragged me by the arms into the darkened grand hall where I was dumped in a heap. Pale moonlight bathed the surroundings in an unreal aura of serenity. My tormentors gathered around -- the thuggish brothers, Eznik, and Omar of Myra. The twins leered at me in anticipation.

"I want you to savor your reward, Caspar," said Omar. "Knowing how you feel about love-making, it would be fitting that you learn to appreciate the love of a husky warrior -- or two -- personally. It may have surprised you that my two friends have shown such restraint in dealing with you. It is merely that anticipation is one of the great pleasures of life, and handling your young body has whetted their appetites and will definitely enhance their enjoyment of you."

Icy terror swept over me, and I tried to drive the images from my mind. Bound like a piece of carpet, I was totally unable to move. My mouth was sealed by the thick gag.

"But this evening shall not be the final act, the *pièce de résistance*, as your French friends say. In several days, after we have tired of you, Caspar, I shall turn you over to your British officers. I shall tell them you are a deserter and a murderer whom I have taken prisoner, and thus will I win their confidence. Because your master, the infidel Flemming, is now dead, I will insure that **you** are blamed. I know how the British treat their deserters, Caspar. I shall have the pleasure of seeing you whipped till you are almost dead. Then I shall relish even more the sight of you being dragged out by your own people to be shot like a dog as a murderer. Thus, you see, not only will your ordeal provide me with great personal pleasure, but it will be the means of my rise to a position of power -- an achievement I'm sure you can appreciate." It had been a well rehearsed lecture, and he had recited it with few of the mumblings which now normally punctuated his speech.

"Unlike you, Caspar, I have made use of the time granted me since your shot changed my life. You have achieved nothing! The lowest rank in an army of slaves! I, however, have been forced to employ talents which had lain dormant behind the beauty you destroyed. In one way I suppose I owe you a debt of gratitude, for without you I would never have become a Bey."

Disbelief must have been written on my face, for he strutted as he went on. "You are surprised? There have been many changes -- many deaths -- since you posed as one of us, Caspar. Being no longer diverted by the pursuit of physical joy, I have risen through the ranks of the Mamelukes. Yes, Caspar, within weeks I shall be awarded the position of Bey now held by an enemy whom I have decided to eliminate. While you are being dragged out to be executed by your countryman I shall be donning the robes of a Bey."

Eznik cut the filthy gag from my mouth. As I gasped for air he whispered in my ear, "The better to hear your screams for mercy, Caspar".

"Why do you help the French now?" I wheezed, trying to ignore Eznik. Much better to hear Omar's boasting and delay what must follow. "You once hated the French."

A SECRET OF THE SPHINX

"Power, Caspar, Power. The French thought to stay forever and usurp our rights. Now the British have replaced the French, but we Mamelukes shall outwit them as well. They have already guaranteed to protect us from the Turks. Under their protection we are consolidating our power. In fact, south of Cairo everything is controlled by us. When the time is ripe, like a cobra in a sack we shall strike the British, and **all** of Egypt will be ours. Once we Mamelukes have regained our proper role, *I* shall rise to rule Egypt. I have seen my Kismet. My night of agony on the Sphinx revealed the path to domination. I have already employed my power to control the fate of lesser creatures like yourself and that intruder, Flemming. Your Kismet controls you, but *I* control your Kismet. You are but a pawn in my hand, Caspar."

"My comrades will be searching for me," I asserted, more to convince myself than for any other reason.

"Only to take you back in chains as a deserter," laughed Omar out of the mobile side of his face.

Desperately I tried to keep him talking to delay the inevitable horror. "I have friends who will protect me. Officers who will stand by me -- Colonel Cameron, Smit' Bey, Mister Flemming. You didn't killed Flemming! He's too smart for you."

"Regrettably Flemming did avoid the hooded friend that waited for him in his lodging upstairs. But he cannot have avoided his fate with its angry mate. His weakness was easy to exploit. Flemming's desire to find the black stone enabled me -- through my faithful *seradj*, Eznik," and here one side of his face smiled tenderly at the pretty little wretch -- "to lay a trail that led Flemming inevitably to the fatal sarcophagus and his Kismet."

Strangely, I felt a wave of comfort. The bastard had seemed almost larger than life, a monster, but now I knew he was not infallible. I would suffer mightily, but I would not accept his will to control me. The Sphinx was right.

"It is an irony that your humiliation will commence here in the Institute," he continued, indicating the surroundings with one hand, "the place where I watched you labor so long in vain looking for the black tablet. How futile your search was! The French who used to live here took it with them when they fled. Foolish infidels, to steal such a thing! The stone is of no value. The French *moallems* carried it with them when they sailed from Alexandria, but twice you foolish British captured them and sent them back -- with the stone hidden in their ship, under your very noses! Infidels are such fools, but the most ridiculous of all are you British. Even the French are able to trick you! Now the worthless black stone is in Alexandria."

He turned to Eznik. "Go to the room used by Caspar and bring his uniform from its hiding place. We will have need of it later." The smaller man murmured something and disappeared. "Now, my friends," Omar addressed the twins, "it is time for you to enjoy your reward. Tie him to the table!" With that the ugly pair seized me and threw me down with such force that the air was knocked out of my lungs. I was stiff from my bonds, weak from lack of food, and so winded from my handling that I'm ashamed to admit I put up a disgracefully feeble resistance. In moments I was naked, bent over a table with my hands tied to each side and my feet on the floor.

This was possibly the worst moment of my young life. Never have I felt so degraded and hopeless. My head was free to move, and I watched helplessly as the thugs stooped to tie my ankles to the legs of the table. When they had finished one of the brutes stepped forward, a salacious leer upon his ugly face. He was naked from the waist down! I withstood the temptation to scream in terror and simply closed my eyes.

"Now is the beginning of a very long night of reckoning for you, Caspar," came Omar of Myra's muffled but satisfied voice. "Begin!"

BLAM! The crash of a Brown Bess reverberated off the walls, and I opened my eyes and looked over my shoulder in time to see the thug fly backwards. As he did, a stain of blood spread over his shirt. I turned my head, and there in front of the table, emerging into the moonlight was Taddy Fromm. He dropped the smoking musket and unslung another from his shoulder. Taddy had found my musket.

"You alright, Yank?" he asked almost nonchalantly.

"Thank God, Taddy!" was all I could manage.

"You speak their lingo, mate. Tell them to untie you."

I turned my head and saw Omar and his man glaring beyond me. I ordered them to untie my feet. Sullenly, they bent below my sight and I felt the bonds loosen. I moved both feet to make certain. "Now Yank, tell the big bastard to come around 'ere and untie your --" Taddy's voice was cut off by a gurgling sound, and I heard the musket clatter to the floor. I twisted my head in time to see a knife at his throat as a red line materialized and burst into a spout of blood. Taddy grasped at his neck, and his mouth moved wordlessly. Behind him loomed the solemn face of Eznik. As my mate collapsed I saw the despair in his eyes.

I heard sobbing! I turned my head over my shoulder to see the remaining brute blubbering and crying out, "My brother! My brother!" He put a hand over his face and uttered a cry of rage, then lunged towards me brandishing a knife. The towering thug poised over me, his hatred over-powering his lust. He had gone mad! A pistol roared and he collapsed, face-down on my back. He slid slowly down my bare buttocks onto the floor.

Bloody dagger still in hand, Eznik gasped in astonishment. With a screech he leapt towards me. I strained against my bonds, but in vain. Eznik raised his dagger above me, leering triumphantly. Suddenly a bolt of light seemed to shoot from his heart. Eznik staggered backwards, a glittering sword-blade projecting from his chest. The unseen swordsman withdrew the blade and Eznik collapsed revealing Mehemet Ali. His other hand held a pistol from the barrel of which a wisp of smoke still eddied. He scowled about the garden, which had suddenly grown silent in the moonlight.

"Where's Omar of Myra?" I croaked.

"You mean the Mameluke? He seems to have vanished. Never mind, my friend, the Mamelukes are finished," replied Mehemet Ali. He looked down at my uniform which lay in a pile at Eznik's feet. Poking the red jacket with his sword, he asked in surprise, "You are a mere Janissary? -- a Private?"

"I'm a Piper," I replied.

Mehemet Ali walked slowly around the table till he stood behind me. His sword slammed into its scabbard and he gazed at my buttocks. Suddenly he drew a knife from his belt, stepped in front of me, and without blinking cut one hand loose. As I scrambled to slide off the table and untie the other, his voice came to me from the darkness, "Now we are even -- Piper."

~ A SECRET OF THE SPHINX ~

I made my way back to the 79th taking with me the corpse of Private Taddy 'Smith'. Taddy would have been amused by the military funeral he received with me playing "***The Lament for the Children***" at his graveside somewhere beside the Nile. Colonel Cameron, I'm sure, had his suspicions about my story of Taddy remaining in Cairo to rescue me, but as Father of the Regiment, he preferred that to the rumor of Taddy's ill-fated palatial brothel.

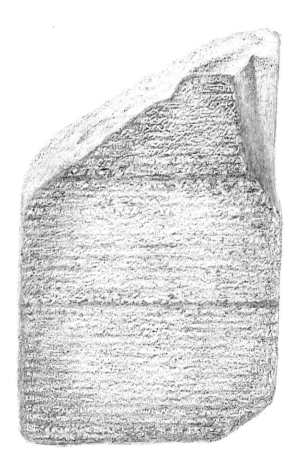

Of course, I advised the Colonel that the Rosetta Stone was in the hands of General Menou in Alexandria, and the rest, as they say, is history. Our army captured Alexandria without much of a fight. Menou signed surrender terms almost identical to those arranged several thousand lives earlier by Sir Sidney Smith. This time the British government did not repudiate the agreement. Although General Menou wriggled and wrangled with Hely-Hutchinson over possession of the Rosetta Stone, he finally gave in with his customary lack of grace. Consequently, the Rosetta Stone has since wound up in a place of honor in the British Museum in London. Maybe Geoffroy Saint-Hilaire will be right. Someone might actually decipher its message some day.

These momentous events were related to me by others. Having rejoined the Regiment, I was once more relegated to insignificant duties such as piping Reveille or Lights Out, or laboring on endless fatigues and loading parties. Had my Vision been wrong? Would the Great Snowy Owl ever glide through my life again? Or was it my Kismet to forever endure the boredom and harshness of a soldier's life in some remote station all the while reliving the tragic memories of so many dear friends and of a love lost?

CHAPTER 17
"A Last Tune in Egypt"
(Alexandria, 20 & 21 October)

Finally the last French troops having embarked for their return to France, the 79th, along with the rest of our little army boarded ships bound for new stations. We Camerons were a bit blue over our pedestrian role in liberating Egypt. We had performed resolutely in a strenuous venture, but had won none of the glory we had hoped for. Of course, we all recognized that the 79th was bound for fame. Nevertheless there was an air of frustration amongst the Jocks as the last of us still ashore packed our personal kit and dismantled the tents.

"MacHugh, front and center!" came Corporal MacAnch's voice, in a roar that brooked no delay.

Crashing to 'Attention', I bellowed, "Corporal!" and waited for a dressing-down for some infraction of His Majesty's rules of conduct. The last few weeks had seen all the departing regiments return to peacetime standards of drill and discipline, which meant constant harassment by NCO's. Corporal MacAnch was bucking for Sergeant, and although he was a good fellow, he wasn't going to let any breach of "discipline and good order" go unnoticed.

"Report to the Colonel's tent on the double -- with your pipes." When my amazement showed he bellowed, "On the double, ye idle wee man!"

Within moments I presented myself to Old *Ciamar A Tha Thu's* orderly, praying fervently that my clobber was up to snuff despite the lack of preparation time. I was ushered inside and was both surprised and delighted to find Mister Flemming. "MacHugh, it is indeed good to see you again. Colonel Cameron was just telling me of his adventure -- and yours -- in the tomb of the Great Pyramid. I understand you also had a busy time in Cairo after I left. The rumours are juicy, but I will not ask for fear of being disappointed by a mundane truth."

"Thank you, sir. Your delicacy is much appreciated," I replied, marveling at my own boldness in speaking so confidently to a gentleman and a Colonel.

"Fate has drawn the three of us together one more time," observed the Colonel genially. I was astounded when he passed me a glass of whiskey. "To the officer, the antiquarian, and the piper," he intoned, "May there be more such encounters." Silently we raised our glasses. Old *Ciamar A Tha Thu* was a true democrat. No wonder we were all willing to die for him.

"To our last night in Egypt," intoned Mister Flemming, and again we sipped. "I have prevailed upon Colonel Cameron to loan your vaunted piping skills to the cause of international harmony tonight, MacHugh."

"Aye, MacHugh, much as I fear my breechless bairns being corrupted by fly-by-night civilians, tonight I am entrusting you to one of the very same. You're a canny lad, and will no be led astray or taken in. Will you, MacHugh?" he bellowed loud enough for the entire Regiment to hear.

"Yes, sir," I replied, wondering what he meant and trying to stand ramrod-straight while holding my dram.

"Relax!" he roared, then catching the humor of it, added, "That's an order!" He laughed and slapped me so hard on the shoulder that I splashed a drop of his good scotch whiskey on my kilt. Then softly he added, "Relax, laddie, we've no formalities here. In a few minutes you will be awa' with Mister Flemming to Aboukir to serenade as sly a pack of rascals as you'll find anywhere in the world -- Turkish Pashas and Mameluke Beys. Is that not so, Flemming? After that -- who knows when we will meet again?" He gave me an earnest and penetrating look.

Mister Flemming then filled me in on my task, something of an honor, it appeared. The senior Turk, a naval officer titled the *Capitan Pasha*, was that night hosting a banquet aboard his flagship to honor the leading Mamelukes. Mister Flemming was to represent the British government, and I was to add a touch of British martial pomp to the proceedings by piping the honored guests along the jetty to the flagship.

"There's just time to have a last tune in Egypt," observed the Colonel. "Jamie, there's few can play **'The Camerons' Gathering'** like MacHugh here." (He used the English title for the benefit of his guest.) "This tune is a favorite of mine, and one to fill even a Sassenach's heart with pride. Go ahead, lad, give us your best rendering." I did so, and felt very satisfied with myself at the conclusion of the old *piobaireachd*. Finishing off with a flourish, I turned to the Colonel, and was surprised to glimpse a tear in his eye. We all knew he was sentimental, but -- ? "I trust we will all live to share this tune again some day. Watch your step, laddie. You'll be consorting with many a sly devil," and he nodded towards Flemming. Then the Colonel did a surprising thing, he grasped my hand and shook it. "Remember, laddie, you will always have me at your back."

Outside Flemming added almost off-handedly, "Oh yes, MacHugh, bring your kit -- and arm yourself -- dirk and pistol -- nothing too obvious though."

A few minutes later, bearing my knapsack, I boarded a carriage with Mister Flemming, and soon we were bouncing along the road to Aboukir. "You have just heard the official version of our outing, MacHugh, now I will fill in a few of the essentials. Some time ago General Hely-Hutchinson, made an agreement with the Mamelukes because at the time we needed their cavalry. In return we would protect the Mamelukes and even support their attempt to take control of Egypt when we leave. The Turks, as you can imagine, were furious when they discovered the agreement. I predict a bloodbath tomorrow, the moment the last British ship pulls out of Alexandria harbor. Others, higher up, share my apprehension, but dare not intervene. It will be our job -- yours and mine -- to prevent the inevitable. Does that sound daunting, MacHugh? Believe me, it is. I wouldn't trust the *Capitan Pasha* as far as I can spit a cannonball. He has invited all the Beys in northern Egypt to a banquet in their honor, **and** they have accepted. We don't know what is really going on, but you and I are to discover what it is **and** prevent it. Old Hely-Hutchinson just wants to get away with his army before things fall to pieces here. You and I are deemed expendable to those in command, and I believe wily old Allan Cameron has sensed that. However, I have no intention of being 'expended', as the tars say, and no doubt a youngster like yourself has plans for the future, so let's do this job and survive. Besides, if we get out of this operation, I have been authorized to make an offer of employment to you that will take you far from the 79th, and lead to a much higher status than that of Private -- or even 'Piper'. What do you think, MacHugh? Care for a little adventure and romance, and at the same time see the world?"

"Yes, of course, sir I don't believe I was cut out to be a peacetime soldier."

"Excellent! Now all we have to do is survive the next twenty-four hours."

~

'At last! Here they come,' I said to myself. Resplendent in their gaudy *yaleks* and *cahouks*, a score or so of Mameluke Beys swaggered onto the darkened jetty. They had arrived much later than scheduled. Typical of their insolence, I decided. Maintaining a dignified, almost menacing silence, suspicion was written plainly on their dark, mustachioed faces. Their leader raised his hand on sighting Flemming who leaned nonchalantly against the railing accompanied by two Turkish naval officers. When all three stepped forward into the torch-lit circle, I shouldered my pipes and prepared to strike up. Aboukir was blanketed in darkness. The only sound was the lapping of wavelets against the shore and the pilings.

The Turkish officers smiled, and one launched into a short speech of welcome in the name of the Sultan, assuring the Beys of his sincere admiration and that of the Ottoman Navy towards all "heroes" of the recent victory. The Mamelukes merely scowled in reply. Next Flemming spoke, claiming to represent the King of England and the British government, and added a nice touch about "enduring peace among the victors of the recent war to liberate Egypt from the French oppressors". The second Turk then invited the guests to follow them to the end of the jetty where they would board the Admiral's flagship to begin the festivities.

This was my cue. I struck up the pipes and stepped behind the two Turks, Flemming, and the leading Bey. I had considered opening with a tune I had just composed in the coach and had played for the first time an hour ago using my need for a "warm-up" as an excuse to try it aloud. Actually, I liked this bouncy little melody I had discovered, but it was a wee bit too informal for such a dignified occasion. That fact is also why I decided to name the tune after Old *Ciamar A Tha Thu.* As it worked out, I never did get to play it again that day. At the last minute Mister Flemming warned me to watch for trouble, so I decided to play **Highland Laddie**, a really easy tune, and keep my eyes and brain open. Along the quay I strode, my polished brogues echoing above the shuffling of soft Mameluke boots. When we were fifty yards from the end of the jetty and the towering side of the flagship, a large body of *janissaries,* the Honor Guard, ran down the gangplanks and formed up facing us. They presented arms. That must have been the signal.

As I piped jauntily along, the two Turkish officers in front of me suddenly turned on the Bey between them and stabbed him repeatedly with *yataghans* drawn from beneath their robes. Then one wheeled, and with the point of his short sword at Flemming's throat, forced him to back away. I ceased piping, somehow got the pipes under my arm and drew the dirk from my belt. Seeing me armed, the Turk indicated he would slash Flemming's throat if I came closer.

"MacHugh! Stay back. They won't dare harm us," Flemming yelled over his shoulder as he raised his hands in surrender. It appeared he was right, for the two murderers took off towards the advancing "Honor Guard". From behind me came screams of rage and the metallic swish of scimitars being drawn. Standing back to back with Flemming I watched the Turks advance on the trapped Mamelukes. An order rang out and they halted, raised their muskets, and prepared to fire. When the officer commanding them drew in the breath to utter the next fatal command I yanked Flemming down onto the planking. The order rang out as our knees hit the jetty, and by the time they fired we were flat on our bellies. The roar was followed by the hissing and whining of musket balls overhead, then screams of pain and rage from behind us. Some of the badly aimed shots tore into the planks only a few feet away while others seemed directed at the moon a quarter of its way across the heavens. "No wonder they couldn't stop the Frog-eaters!" came Flemming's disgusted voice.

He leapt to his feet, and despite being unarmed, stepped between the Turks and their victims. Dirk in hand I followed his example. Ringed by fierce *janissaries* closing upon us in silence, I heard Flemming beside me. "MacHugh, do **not** attack or fire that blunderbuss of yours. That is an order!" It dawned on me that a small detachment of Turks were paying all their attentions to the two of us, shepherding us across the wharf until our backs were against the railing. Heedless of the screaming and clash of arms around them, they stared stolidly at us as the slaughter continued without them. Protecting us was their sole duty. Realizing this, I spared a glance towards the unfortunate Mamelukes -- my former comrades. Despite the initial volley and the overwhelming number of their assailants, they were making a great stand. If it hadn't been for the tightness of the area in which they were confined, I believe they would have slashed their way out although another party of *janissaries* now blocked the pier behind them. A second volley rang out, and more Mamelukes went down. Their only hope was to leap into the water and swim. A few tried that, but the crack of muskets below and the sound of oars proved the *Capitan Pasha* had prepared for that eventuality.

Aware that several Beys had taken refuge behind Flemming and me, I placed myself to protect them as well as I could. Our escort of *janissaries* made no move to come closer. Now only half a dozen of those magnificent Mameluke swordsmen stood, wielding giant scimitars in each hand, killing all bold enough to advance against them. Another ragged volley roared. Except for the groans of the mutilated and dying, silence descended. Now there were but four Mamelukes, surrounded by the corpses of their comrades and their enemies. Still their scimitars flashed in the torchlight, though no enemy stood within ten feet of them. What a magnificent sight they were! If ever I was proud to have been a Mameluke, it was that moment!

A single shot rang out and one of the four fell. There was a moment of silence, then a fusillade erupted, and none remained standing.

"Flemming Bey, it is unfortunate that you were present to witness this," came a voice in passable English, "but you of course saw clearly that these Beys resisted arrest -- an arrest order from the *Capitan Pasha* himself." It was one of the two original murderers speaking. "I request that you step aside while we arrest the three criminals who have taken refuge behind you."

"I shall defend them with my life," declared Flemming. "If pressed, I will acknowledge that the others resisted your attempt to arrest them. However, these three have surrendered to me, the representative of the British Government and King George. I will not relinquish them. They are now under the protection of the Royal Navy, and any harm that befalls them -- or me -- will result in the destruction of the Sultan's fleet. Lord Hely-Hutchinson and Admiral Keith know of my mission. If I do not report to them by morning, the destruction of your entire fleet will be the result."

My heart swelled with pride as I listened to his words. If I could only be so courageous, standing unarmed before an overwhelming mass of murderers, employing my wits to defeat them. I was now sure Flemming had come here on his own hook. There were no orders to destroy the Sultan's fleet. The bigwigs back in Alexandria were only interested in getting out of Egypt. I could hear the shuffling of the three Mamelukes behind us, and supposed they were wondering when they would die.

The *Capitan Pasha* himself turned up at that moment. Already the dead of both sides were being dragged away, the passions of the last few moments replaced by the mundane task of tidying up. He spoke to Flemming reassuringly. "Flemming Bey, I would be most honored if you and your prisoners would join me for tea." Apparently I had become invisible. "I am satisfied to leave them in your care. I would be very much distressed if you were to feel at all threatened or alarmed by these unfortunate events." The *Capitan Pasha* then ordered his *janissaries* to step back and ground their weapons. They did so, and we all relaxed marginally. The grand Turk advanced wearing his sincerest artificial smile and took Flemming by the hand. They shook, and the crisis ended. It was only then I realized I still held my pipes at the 'Ready' position under my left arm. I would have certainly put up a deadly fight!

"I can put your prisoners under my protection if you like," offered the *Capitan Pasha*. "Otherwise, you will have to guard all three with only the assistance of this musician. You have my word that nothing will happen to them. For after all, they are your prisoners, not mine."

While the two discussed the issue, I turned to inspect the prisoners I might have to guard. The foremost was a huge, fierce-visaged devil who looked as if, given a second chance, he would have preferred to die with the others. His neighbour was a shifty-eyed wretch. Behind them, all in black, stood the third, his face also swathed in black so that only his eyes glinted in the flickering light. Full of hate, they were fixed upon me! Shocked, I returned his stare. Then deliberately he reached up and unwound the fabric covering his face as he turned slightly to reveal the shattered cheek and the baleful sagging eye. Omar of Myra!

I shuddered, which I suppose he mistook for fear, but within seconds he no doubt recognized the hatred in my stare. I owed him for so much, but most of all for Zenab. Silently I vowed to kill him, and stepped forward a pace brandishing my dirk, when a hand was placed on my arm. "MacHugh, come out of it, man!" Flemming's voice was angry. "Pay attention!" he ordered in French. Reluctantly I took my eyes off Omar of Myra and paid full attention to Flemming. "Now listen, and don't say a thing. This bloody Turk doesn't understand French. I have convinced him to let you return to Alexandria, supposedly to prevent our fleet from coming to destroy his fleet. That is our ace. But there is no such order. So **you** are the Joker. You must convince someone there to come and rescue me and these survivors. Find Colonel Cameron. If anyone will believe you and wake the dunderheads it is he. You must convince them to bring a show of strength here at first light or we'll have abandoned Egypt to civil war. Understand?" I nodded, but was distracted by the sight of our three 'prisoners' being disarmed and marched away by the Turks. "I will arrange to have a horse for you," he concluded.

Flemming turned to speak with the *Capitan Pasha,* but found himself already dismissed and in the hands of the Second-in-Command. Consequently, it took almost ten minutes to make arrangements. Finally I had an a opportunity to speak to him in private. "Sir, did you see your three so-called prisoners?"

"Not really," he replied, his mind on other things.

"You really should look them over. The ugly fellow in black is Omar of Myra, the murderer who made the attempt on your life in Cairo. Don't accept any gifts from him, sir, especially if they come in a sack that wriggles. Here, you had better keep my pistol at your side," and I passed him Colonel Phelippeaux's bell-mouth St. Etienne. "See you tomorrow."

I found the Mamelukes' horses after a search. The *serradjs* who had been in attendance on the Beys had also vanished, although traces of blood remained at the end of the jetty. The horses had been tethered and left unguarded a short distance away and now stood patiently waiting for masters who would never return. I chose one that looked like it could move, and was about to untie it when I realized I still held my pipes under one arm. Habit is an amazing thing. I found a saddlebag large enough to carry them and knelt to stuff them into the bag. A step sounded behind me.

"Ah, I have found you at last, Caspar." It was Omar of Myra. "It pained me beyond belief when I lost my chance to exact revenge upon you that night in the Institute. Allah has indeed been kind to restore you to my power." He spoke softly, but not from fear of detection. The moon illuminated his two faces. Both displayed complete confidence. "I have done as my Kismet decreed, infidel. While you have sunk to the bottom, a common soldier in an army of slaves, I have become a Bey -- now one of very few. My rivals lie in their gore. Do you know why? Of course you don't. It was I who convinced them to attend the *Capitan Pasha's* banquet. In fact, the banquet was my plan. The foolish Turk could not see, let alone understand, the scope of my subterfuge when I suggested luring them to him with such a simple ploy. At this minute, in Cairo the remaining Beys are being rounded up and destroyed. The Turks believe they have won, that they have destroyed the edifice of Mameluke supremacy, but they have, in reality, only disposed of the rubble, leaving the way clear for me. I shall unite the remaining Mamelukes into one indomitable army which in due course will see the Turks driven from Egypt as were the French and now the British. Ironic, is it not, that I owe all this to you, Caspar?"

I sensed he wanted to boast, **and** I needed to learn more, particularly about the fate of Flemming, so I finished packing the pipes, and asked innocently, "What makes you think the

Capitan Pasha won't hunt you down and kill you once he learns of your escape?"

"I did not 'escape'," he retorted as though I were a fool. "He thinks I am his tame Mameluke, one who works for him in return for a promise of wealth and safety. None realize it is I who control all their destinies. At this moment the *Capitan Pasha* holds Flemming and the two surviving Beys -- hostages -- in case the British react to tonight's cleansing. He 'ordered' me to prevent you from returning to Alexandria, but the fool does not understand that I take no orders from him, and am here solely to avenge myself."

"But by letting me reach Alexandria," I pointed out, "the *Capitan Pasha* will save his fleet from destruction by our navy. Why would he court disaster by preventing me from calling them off?"

"You British are as great fools as the French and the Turks," he scoffed. "I speak and understand French, of course. My time in El Bekri's service was not wasted. I heard Flemming tell you there is no plan to destroy the fleet. I told the *Capitan Pasha* so, and that is why he agreed that you must die, Caspar. Of course, when I goad him into another foolish act -- the murder of Flemming -- the British **will** destroy his fleet, so I won't have to do so later." Omar of Myra drew himself up proudly, convinced he had thought of everything.

Still crouched over the bundled bagpipe, I whipped off my tall feathered-bonnet, threw it in his face, and ducked backwards beneath one of the horses. I had to stay in a confined space. Otherwise Omar's two whirling scimitars would make short work of me. Armed with only a dirk and the tiny *skean duhh* hidden in my stocking, I was no match for an angry

Mameluke in the open. Omar of Myra dashed after me into the maze of horses. As we ducked and weaved in and out among them, they grew more restive and side-stepped or pulled on their tethers. From behind one I taunted him. "Omar of Myra is good at getting others to kill for him, but not so good at doing the job himself." Then I offered a momentary target for his rage -- my insolent face. He swung awkwardly across the broad saddle with one of his swords, but I was no longer there. Seconds later I popped up behind another horse and taunted, "Omar of Myra begs his mother, the cobra, to kill his enemies for him!" Again he swung with the scimitar -- too late. He would catch on soon. It was time to risk all.

I slipped behind another mount and sneered, "Omar of Myra, killer of girls!" I nearly choked as I said it, but paused a split second more. He could not resist the bait. As his blow slashed the darkness I ducked under the horse and stabbed him in the thigh with my dirk. He screamed in disbelief; his Kismet had lied! I was upon him, dirk in one hand, *skean dubh* in the other. "This is for Zenab!" I cried and dirked him in the stomach. He screamed in agony and terror. "This is for Taddy!" and I slashed his throat with the deadly little black knife. Omar of Myra clutched at his throat, slumped to the stable's filthy floor, and gurgled for a moment. Then all was silent.

~

"Colonel Cameron, sir, wake up, sir!" I shouted in the giant's ear. He leapt up, nearly bowling me over. It took only seconds to tell him Flemming's message, and minutes before he presented me to General Hely-Hutchinson. Within an hour troops were being disembarked, among them a party of the 79th.

After a short forced march we strode into Aboukir with me at the head of the 79th piping my second-to-last last tune in Egypt, *"The Pibroch of Donald Dhu"*, jaunty as can be. There was still time for another short tune so I yielded to temptation and tacked a wee new tune on the end. As a result the 79th strode the last few yards to the strains of *"Our Old Ciamar A Tha Thu."* I thought I detected a grin on the giant's face as we wheeled into position and came to a halt. There we surrounded the *Capitan Pasha's* tent, and our officers threatened to sink the entire Turkish fleet unless Flemming and the Mamelukes were surrendered. My friend and the two surviving Beys were immediately delivered unharmed, and I was taken along with several officers to identify a row of reeking corpses. The buzzing of thousands of flies almost drowned the subdued murmur of the officers as they reluctantly examined the distorted faces in an attempt to name each victim in preparation for the military funeral being arranged.[58] Eventually there remained unidentified only one bloated corpse. Although Flemming stared quizzically at me, I remained silent as the body was dumped into an unmarked grave. But I had recognized the shattered cheek and that one sunken, sightless eye.

~

THE END

Editor's Postscript

Years after these events, Louis de Bourrienne, Bonaparte's secretary, summed up Napoleon's great Egyptian fiasco: "And what were the results of that memorable expedition? The destruction of one of our finest armies, the loss of the best of our generals, the utter destruction of our navy, the loss of Malta, and the complete domination of the Mediterranean by the English. And what remains of all that today? A scientific work." [Bourrienne, Louis Antoine Fauvelet de, ***Mémoires de M de Bourrienne sur Napoleon***. Paris, 1899-1900, vol. 4, p. 55.]

It may be of interest to readers to know what happened to some of the personalities encountered by MacHugh in this volume of his Memoirs. Unfortunately, many of these individuals have appeared in no other recorded sources, and only by MacHugh's words do we know them. Others went on to fame or notoriety. It is these whom we include here in the order of their appearance.

∽

Sir William Sidney Smith: The talented but exasperating Smith continued his long and illustrious naval career, involving himself in many unlikely schemes while continuing to earn fame and promotion. He was finally made an Admiral in 1821. Sir Sidney retired to Paris and died there in 1840.

"Certainly one of Britain's more colorful naval commanders of the Napoleonic era; vainglorious, relentlessly self-assertive, and constantly ready to exceed his authority, Smith was also daring, energetic, willing to take counsel from more experienced men, generous, and even-tempered; he saw himself as a chevalier and acted accordingly." [*The Harper Encyclopedia of Military Biography*, Castle, Edison, N.J., 1995; p.698]

In 1999 Elizabeth Sparrow's superb *Secret Service: British Agents in France 1792-1815* [Woodbridge, 1999] revealed many of the secrets of Sir Sidney Smith's remarkable career. Much of the time Smith had actually been serving two masters, The Royal Navy and The Secret Service. Hence the repeated conflicts with his naval superiors. Even his strange capture and subsequent imprisonment in The Temple in Paris which culminated in his famous escape, are revealed to have been staged by Smith and the Secret Service to enable him to set up and control a spy network in the enemy capital. Sparrow's diligent efforts have revealed Sir Sidney Smith to be an even more remarkable character than his contemporaries dreamed. For instance it is now known that Smith had planned to prevent Napoleon from seeing the newspapers at Aboukir in August, 1799. The plan was to steer Bonaparte towards Italy, the scene of his greatest successes, without revealing that the Army of *le premier l'homme de l'Europe* no longer existed. Sir Sidney later told his brother that his intention was to capture Napoleon at sea.

Lieutenant Fourès: Little is known of the further career of the Gascon. He did resign his commission and returned safely to France to vanish from the prying eyes of historians.

A SECRET OF THE SPHINX

Roustam Raza: The Mameluke served as Bonaparte's bodyguard, valet, and procurer for many years to come. In his exotic garb he became almost as famous as the Emperor himself, and was often immortalized in paintings depicting him astride a prancing Arabian. Influence-peddling and extortion (skills natural to a Mameluke) enabled Roustam to amass a sizeable fortune. Sensing the fate about to befall his master, Roustam deserted Bonaparte shortly before the Emperor's abdication in 1814, then married and wrote his memoirs -- "which reveal a naïve, illiterate, shrewd, and bullying lackey." [Herold, p.326] No mention of MacHugh or either of his aliases, Gaspar Genereaux and "Caspar", is made in these memoirs (***Souvenirs de Roustam, mamelouck de Napoleon Ier***, Paris, 1911.)

Pauline "Bellilotte" Fourès: Mme. Fourès, who after her divorce styled herself '*Mademoiselle* Bellisle', after an alleged affair with General Kleber, managed on her second attempt to return to France in 1800. There she married twice, became a novelist, harpist, painter -- and a noted eccentric. She lived in exile outside Paris supported generously through the years by Bonaparte. After his downfall she prospered as an importer of wood from Brazil. "Bellilotte" outlived all MacHugh's acquaintances, dying in 1869 at the age of 89 shortly after the opening of the Suez Canal by French bankers and engineers, and just before the collapse of another French Empire under another Napoleon Bonaparte.

THE MacHUGH MEMOIRS ~ (1798 - 1801)

General Napoleon Bonaparte: After deserting his "Army of the Orient", Bonaparte returned to France, surprisingly enough, to a hero's welcome. As one of a triumvirate, he seized power in a coup d'état less than three months later, on 9 November, 1798. On 18 May, 1804, he crowned himself Emperor, and ruled France and much of Europe until defeated and forced to abdicate by an allied coalition in 1814. He returned in 1815 for "The Hundred Days" which ended with his total defeat in one of the most famous battles of all time -- Waterloo. Napoleon Bonaparte died in exile on St. Helena in 1821.

~

Lieutenant John Wesley Wright: The Lieutenant survived his wounds from the sortie at Acre and served with distinction. In 1803 he was entrusted with secret dispatches to the British Embassy in Paris. In command of ***H.M.S. El Vincejo***, Captain Wright was captured again in 1803 and was imprisoned in the same room in The Temple Prison in Paris from which he had escaped in 1798. He was murdered in his cell in 1805. The case has never been solved.

~

Eugène de Beauharnais: "In 1804 he was made a prince of France, and in 1805 a viceroy of Italy. In 1806 he married Princess Augusta Amelia, daughter of King Maximilian I of Bavaria, and was formally adopted by Napoleon and made heir apparent to the throne of Italy. Honourable and sagacious, he showed great military skill in the campaigns in Italy, Austria, and Russia. After Napoleon's abdication in 1814 he retired to Bavaria, and was created Duke of Leuchtenberg." (**Chambers Biographical Dictionary**, p.122-3)

Colonel Mehemet Ali: Mehemet Ali remained in command of his Albanians in Egypt. In 1805 they mutinied after not being paid by the Sublime Porte. Being the only disciplined force in Egypt, they prevailed and proclaimed Mehemet Ali "Pasha of Egypt". On March 1, 1811, he massacred the Mamelukes in the citadel in Cairo, bringing to an end their centuries of dissolute domination. As 'Mohammed Ali, Khedive of Egypt', he modernized Egypt in many ways, and greatly bettered the lot of the *felaheen*. Known as "The Father of Modern Egypt", he was also active at home, fathering ninety-five children. Almost equally successful as a military commander, he went on to annex parts of Arabia, Nubia, and the Sudan, building most of what is modern Egypt. In 1848 he became insane and died a year later. His family's reign ended rather ingloriously in 1952 with King Farouk.

The Rosetta Stone: The "lump of black basalt" retained its secrets for another few years despite the efforts of several men of genius who each went part of the way towards deciphering the inscription. Finally, in 1824, the French Egyptologist, Jean François Champollion, published the complete translation of its hieroglyphic inscription. It still is one of the most famous exhibits of the British Museum in London.

Etienne Geoffroy Saint-Hilaire: The great zoologist returned to France and the academic life. He became renowned in scientific circles for technical publications and for his theories which foreshadowed Darwin's. Geoffroy became embroiled in a bitter and much publicized battle of the biologists with his friend, Cuvier, "the father of Paleontology". Geoffroy died in 1844.

~

François-Auguste Parseval-Grandmaison: "The Bard of the Nile", as MacHugh called him, was later described by a prominent historian as "a less than mediocre poet, but a man who possibly did not deserve all the ridicule heaped upon him by historians." [Herold, p.31]. Otherwise, Parseval has vanished into well-deserved obscurity.

~

Lieutenant-Colonel Alan Cameron: "Old *Ciamar a tha thu*" or the "great Goth", as a compatriot described him, continued in command of his beloved 79th until promoted to command a brigade at Talavera and Bussaco. In 1810, at the age of 60, an advanced age for an officer on active service, Alan Cameron was invalided home. Lieutenant-General Alan Cameron KCB died in 1828 in London. A window was erected to his memory in the Church of Marylebone in 1948.

~

The Sphinx: The largest surviving sculpture of the ancient world has retained much of its mystery to this day. In 1816 and 1817, when the wars instigated by General Bonaparte had finally ended, the Sphinx became the subject of another in a centuries-old series of excavations to examine its true form. A Genoese merchant named Caviglia managed to clear away some of the sand and discovered fragments of a beard which appears to have been

∽ A SECRET OF THE SPHINX ∽

added to the sculpture after its construction and later removed, the reasons for which are still being debated. These fragments of the beard are now on display at the British Museum in London. Many more excavations have taken place since, each resulting in some new discovery and related mystery. Today pollution and wind-blown sand are the greatest threats to the Sphinx.

Glossary Of British Slang

The following are terms commonly used in England during the early 19th Century unless otherwise noted.

blanket hornpipe: amorous congress, copulation

blowen: prostitute

Bosun: boatswain: the warrant or petty officer in charge of the crew

Brown Bess: the musket supplied to the British Army for nearly a century

clobber: personal equipment and uniform

cove: a man, usually a rogue

dell: a buxom young girl

dial: face

dicey: risky

diver: a pickpocket

dodge: profession, occupation

doodle: a silly fellow

doxie: a prostitute

Farmer George: George III

fence: one who buys and sells stolen goods

fire-ship: a girl infected with venereal disease

flash: ostentatious; also: knowing; also a verb: to display or show

Flash Man: a pimp or a brothel bully

gammon: to deceive

gander: to look

greet: to weep (Scottish)

hopper-arsed: describing a person with a large backside

lobster: a soldier [from the red coat] (naval)

molly: a homosexual man

nob: the boss

peppered: infected with venereal disease

Pipie: Pipe Major (military)

plug: to work

poxy: infected with venereal disease

quid: cash; also: a guinea; also: a mouthful of tobacco

rough it: sleep on the bare floor without removing clothes

rum: good, valuable, excellent

rum duke: a queer unexplainable fellow

Sassenach: an Englishman (Scottish)

screw: a prison guard or warder

shaver: a cheat

shill: a salesman's confederate who pretends to buy to convince others

snaffle: to steal

swadd, swaddie (also **squaddie**)**:** a soldier

swell: a well-dressed man

swiving: fornication

tackle: the male reproductive organs

tar: a British sailor (from "Jack Tar")

top: to kill or finish

weevil-eater: a sailor

windjammer: a sailor

Glossary of Turkish and Egyptian Terms

Ashraf: a person of Noble descent (see also Sherif)

Bey: literally 'lord'. The Beys were Mamelukes who ruled Egypt like feudal princes for their Turkish masters

Cahouk: the Mameluke head-dress, similar to a turban

Capitan Pasha or Pacha: supreme commander of the Turkish fleet.

Caravanserai: a large fortified resting place to accommodate entire caravans overnight

Charoual (or 'Serouel'): the baggy trousers worn by the Mamelukes

Chiftlicks: a famous regiment of Janissaries; They landed at Acre with orders to serve under Sir Sidney Smith, almost as his personal guard

Effendi: Turkish term of respect, the equivalent of 'master'

Emir: Arabic term for Prince, governor, or commander

Fellah (pl. fellaheen): Egyptian peasant

Hadj: the annual pilgrimage to Mecca; title given to one who has returned from the pilgrimage

Harem: the wives or concubines of a Muslim of high standing; also the place where the wives or concubines live. (also '**Seraglio**')

Imams: religious officials who lead the faithful in prayers

Infidel: a non-Moslem

Janissaries: the name given to the Turkish infantry until 1826

Khamsin: the annual strong and hot south wind

Kismet: fate, destiny, one's future

Mamelukes: a military caste that had originated as slaves who seized power in Egypt in 1250, and continued in control until 1517; Since that time they remained powerful although under the control of the Ottoman Turks who ruled the Levant

Moallem: a title signifying something between 'Mister' and 'Doctor' and normally reserved for educated Christians

Nahib al-ashraf: literally "head of the sherifs"; the badge of office was an ermine caftan

Ortah: A unit of the Turkish Army, similar in structure to a Regiment in European armies, but much larger

Pasha (also Pacha): a title given to high-ranking civil or military officials and to governors of provinces

Seraglio: see '**Harem**'

Seraskier: an officer commanding an Army

Serradj: A Mameluke's attendant, usually a foot-soldier

Sheik: Arabic title of respect meaning 'elder', but given regardless of age to any notable; Not a military title

Sherif (pl. Ashraf): person of noble descent

Sublime Porte: the seat of the Ottoman government in Constantinople

Sultan Kebir: literally "Commander-in-Chief", but Bonaparte preferred the grander translation "Great Sultan"

Yalek: the Mameluke's colorful jacket

Yataghan: a short sword favored by the Turks, usually worn tucked in a sash

APPENDIX
The 79th Cameron Highlanders

The 79th was raised in 1793 by Major Allan Cameron of Erracht, one of several regiments recruited in the Highlands of Scotland during the wars against the French. The Eighteenth Century British Army was organized on a much different basis than are modern armies. Loraine MacLean of Dochgarroch describes the regimental system best in **The Raising of the 79th Highlanders** (Society of West Highland & Island Historical Research, 1980).

"In the 18th and early 19th centuries the Army was financed in what now seems an extraordinary way. It was run almost as a collection of business companies, each regiment being one company. The colonel of a regiment, who often never saw it, wanted the best return on his investment. He had had to buy every step of his promotion unless he had been very lucky, and so had everyone below him, from Ensign to Lieutenant Colonel, either in cash or by raising men. The officers ran the regiment/business for his and their own profit. Their pay was neither a living wage, nor even a fair return on the cost of their commissions, yet they were expected to live as 'officers and gentlemen' on an income that did not cover the cost of their uniforms.

A SECRET OF THE SPHINX

All they could do was to hold on and pray for a rich wife and a lucrative war, with its chance of prize money, plunder and promotion. ...

The payment of the rank and file, in theory about £12.13.6. a year, by the time the 'Off Reckonings' had been deducted came to something under £1 a year, but the soldiers seem to have been paid. [Whereas the officers were usually years behind in receiving their pay!]* The whole Army was miserably paid, but so were the 'poor mechanicks' in the towns. Where the soldier scored was in the Army's own unique welfare service in feeding, housing, clothing and caring for its employees and its pensioners ... in a way that had no counterpart in the civilian world."

Notes

1. Beadles were originally minor church officials who were appointed to keep order in religious processions and events. Sometimes accepted as public law-enforcement officers.

2. One notorious hulk, **Warrior**, formerly a famous 74-gun battleship, officially listed as holding 450 prisoners, actually contained 638 as late as 1855. Of these, 400 were admitted to hospital and 38 died. This is but one example described by Harry Mayhew and John Binny in **The Criminal Prisons of London** (London, 1862].

3. MacHugh's references to **Justice** may well be an error of memory, as the first "hulk" ever established was the **Justitia** in 1779, and it is probable that it was still in use in 1798.

4. The criminals of London had their own hierarchy: "Cracksmen" or "Rampsmen" were plunderers by force -- housebreakers and violent criminals. They were considered the elite. "Mobsmen" were next, those who operated in groups and worked the crowds -- various types of pickpockets or "divers". "Sneaksmen" were the lowest class, being the various sneak-thieves.

5. In MacHugh's time the cluster of three buttes located on the Montana side of the border with Canada was known by its Blackfoot name, "The Sweet Pine Hills". A later mistranslation caused this magnificent natural site to be known today as "The Sweet Grass Hills".

6. Hunts With The Owls became the equivalent of a foster-brother to MacHugh in 1797 as related in *"Black War-Bonnet"*, Vol. I of **THE MacHUGH MEMOIRS.** Talks Funny accompanied MacHugh on his departure from Blackfoot country.

7. Sir Sidney Smith had escaped from a French prison only a few months earlier, in May, 1798, and had indeed received his orders to sail and his commission as "joint minister plenipotentiary" on 21 October. [Lord Russell of Liverpool, *Knight of the Sword, The Life and Letters of Admiral Sir William Sidney Smith GCB*, (London, 1964) pp. 64-65]

8. Count Louis-Edmond Le Picard de Phélippeaux (1768 - 1799) had graduated in 1785 with higher honors than Bonaparte, and both began their careers as Lieutenants in the Artillery.

9. "James R. Flemming" is a fictitious name, created by MacHugh to cloak the identity of an operative of the British Secret Service. Although little is known about the man, a "John H. Barnett" did operate in the Mediterranean area and was the known instigator of the "Fourès Affair" [Richard Wilmer Rowan, *The Story of Secret Service*, (New York, 1937) -- 198-201]

10. This identity must have been suggested by young Rory MacHugh himself. 'Genereaux' was his mother's maiden name, while 'Gaspard' had been her favorite name for him.

11. Fourès and his wife, Pauline, are sources of considerable controversy. He is described as a Lieutenant of the **22eme Regiment de Chasseurs à Cheval** by Christopher Herold and his sources [**Bonaparte in Egypt**, (London, 1962)], and as "a Captain of the 20th cavalry regiment" by Juan Cole and his source (Bernoyer) in **Napoleon's Egypt** (New York, 2007). Pauline is described by most "observers" as being a stunningly beautiful lady with exceptionally long blonde hair. Her portrait shows and charmingly cute brunette with short hair.

12. It is generally accepted that Lieutenant Fourès and Napoleon Bonaparte encountered one another in Mme. Fourès' room. However, Juan Cole's excellent book, **Napoleon's Egypt** (New York, 2007, pp. 192 - 196), gives a totally different version in which Lieutenant Fourès did not even leave Egypt, and was never captured by the British. Cole's version is based on Napoleon's letters and the journal of Francois Bernoyer, Napoleon's civilian Quartermaster. Neither source can be considered reliable when dealing with events embarrassing to Bonaparte.

13. On December 1, 1798, Bonaparte had arranged for an unmanned balloon ascent as a propaganda coup. According to some sources it was during this event that

he spotted the beautiful 'Bellilotte' and seduced her the same night. Unfortunately for Bonaparte, his propaganda coup had not been as successful. The balloon caught fire and the gondola fell, scattering Bonaparte's printed proclamations. El-Djabarti, who was an eyewitness to the fiasco, wrote, "The French turned red with shame, for this was not a ship on which, as they pretended, one could travel through the air from one country to another; it was a mere kite." [J. Christopher Herold, **Bonaparte in Egypt**, (London, 1962), pp. 154-5]. According to several sources, Bonaparte first spotted Mme. Pauline Fourès at one of the many balls given by the various French generals prior to the balloon ascent of 1 December, and it took several weeks of persistence by him with the assistance of some of his officers before she succumbed to his "love". [Juan Cole, **Napoleon's Egypt**, pp. 192 -195]

14. "The region of Syria, as understood in 1799, consisted of the present sovereign states of Syria, Lebanon, Israel, and Jordan. It was divided into five provinces or *pashliks*... The so-called Syrian campaign of Bonaparte was fought, not in Syria proper, but in Palestine -- that is, in modern Israel and the Lake Tiberias district of Jordan." [Herold, p.263.]

15. MacHugh's account is verified by several appalled French witnesses such as Major Detroye and Citizen Peyrusse. Although apologists for Bonaparte would later claim that the executions were ordered only after a general council of war failed to find any other solution

because of an alleged shortage of food, J. Christopher Herold, (**Bonaparte in Egypt** p.276) convincingly refutes these specious arguments. "All evidence points to the conclusion that Bonaparte alone gave the order to execute the prisoners, that no one objected or dared to object, and that the execution was carried out with complete efficiency. If there was a reason for the execution, ... it was Bonaparte's deliberate policy to produce a strong impression on Djezzar" [the Commandant of Acre -- Bonaparte's next target].

16. MacHugh is one of very few contemporaries to describe Bonaparte's voice. Karl Schulmeister, the Emperor's most famous spy, also described it as "crisp and strident, rather high pitched, and adds that Napoleon's habit of speaking through his teeth gave a hissing effect to nearly everything he said." [Rowan, pp.235-6]

17. MacHugh was incorrect in this assumption. Although estimates by French officers of the number of victims of the slaughter range between 4,441 (Detroye) and 3,000 (Peyrusse), both noted that Bonaparte spared natives of Egypt (whom he returned to Egypt) as well as 300 French-trained Turkish artillerymen whom he expected would be useful to him. As an Egyptian Mameluke, MacHugh, might have otherwise ended up safely in Egypt.

18. The first French assault took place at 4 a.m. on March 28.

19. MacHugh has related one of the stories behind the name of the most famous tower of Acre -- The

"Damned" or "Cursed" Tower. An older legend states that the tower was so-named "because it was supposed to have been paid for by Judas Iscariot's thirty pieces of silver". [Peter Shankland, **Beware of Heroes** (London, 1975) p.58]

20. The marine was Private David Close. His "dodge" did not last long. Close later stole a diamond ring given Smith by the King of Sweden. The theft was discovered and Close apprehended and kept in confinement for a few days. Then Sir Sidney returned him to normal duties. [Shankland, p.73]

21. Djezzar was said to have eighteen white women and numerous more of other races in his *seraglio*.

22. This was probably General Lannes who fell, wounded by a bullet in the neck, while trying to encourage the main body to follow him through the breach to rescue the doomed first wave. General Rambeaud leading the first wave, was killed, but he does not fit MacHugh's description. Lannes was noted for his size and fair hair. Although MacHugh encountered Lannes later in his career, he does not appear to have made the connection himself -- possibly because of the darkness of the night and the intensity of the action. Lannes was rescued by a Captain who seized him by the leg and dragged his huge body back out of the breach and into the French lines.

23. *"Grognard"* was a slang expression used by the French in referring to a veteran infantryman. It translates roughly as "groaner" or "complainer".

24. Several British officers had been killed that night by Turks who mistook them for Frenchmen. Smith overlooked the incidents in the interests of Allied co-operation.

25. Most other sources conclude there were no survivors from the first wave. Peter Shankland (**Beware of Heroes**, p.81) states that the only survivors from the first wave were a few who had taken refuge in a mosque where they held out until they were able to surrender to Sir Sidney Smith in person.

26. "The historian, La Jonquière, has calculated that French casualties in the Syrian campaign amounted, at the very least, to 1,200 killed by the enemy, 1,000 dead from disease, and 2,300 ill or heavily wounded." [Herold, **Bonaparte in Egypt**, p 299]

27. The wounded were not evacuated because Admiral Perrée had not carried out Bonaparte's instructions, having decided the chances of losing his ships to Sir Sidney Smith's squadron justified his return to France. However, most historians and publicists of the time apparently assumed that an order issued was the same as an order carried out, and so reported that the French wounded had indeed been saved by order of General Bonaparte.

28. General Boyer's detachment left Jaffa on 25 May.

29. Quite a controversy has grown around this event, with Bonaparte's apologists claiming it never took place. But all evidence points to the truth of the story. By direct

order of Bonaparte, Laudanum was definitely administered to the patients by the pharmacist Royer and by Hadj Mustafa, a Turkish physician -- although not, it is suspected, in sufficient doses to kill the patients. Sir Sidney Smith reported that "seven poor wretches" were found alive in the hospital and nursed to health by the Turks. These survivors, along with several other reliable witnesses, verified the story.

30. Murad Bey's famous signal to his wife took place on the night of 13-14 July, 1799.

31. There was no Royal Navy ship, captured or otherwise, named **La Sangsue** (The Leech). Nor did a "Captain Creevey" command any ship in the Mediterranean at that time. MacHugh apparently employed assumed names rather than identify the officer involved.

32. MacHugh's figures agree with Sir Sidney Smith, General Kléber, and Mustafa Pasha. However, most standard reference works still accept Bonaparte's inflated claim without question.

33. Although MacHugh makes no mention of it, Lannes' first wife had been carrying on an affair while he was in Egypt. On his return Lannes divorced her and remarried. His second wife would become a close confidant of the future Empress Marie Louise.

34. MacHugh did indeed correctly identify the military figures who accompanied Bonaparte. They were Generals Louis-Alexandre Berthier (1753-1815), Antoine-François Andréossy (1761-1828), Jean Lannes

(1769-1809), Jean-Baptiste Bessières (actually still a Colonel) (1768-1813), and two who joined the entourage at Alexandria -- Auguste-Frédéric-Louis Viesse de Marmont (1774-1852) and Joachim Murat (1767-1815). Also of the party were Bonaparte's four surviving aides-de-camp; Duroc, Merlin, Lavalette, and his step-son, Eugène Beauharnais.

35. The three members of The Institute of Egypt who were included in the voyage home were Gaspard Monge (1746-1818) (science, mathematics), Claude-Louis Berthollet (1749-1822) (chemistry), and Dominique Vivant Denon (1747-1825) (art). Other civilians included Bonaparte's valet, Roustam Raza; his secretary Louis de Bourrienne; and his cook.

36. This incident is also related with only slight differences by Edmé-François Jomard, [**Souvenirs sur Gaspard Monge**, Paris, 1853, p54.]

37. MacHugh is here referring to the Blackfoot campsite mentioned in **"Black War-Bonnet"**, Vol. I of **THE MacHUGH MEMOIRS**. The site is known today as "Writing on Stone Provincial Park" and is situated on the Milk River in southern Alberta, Canada.

38. MacHugh's strange account of Parseval's arrival and boarding of **La Muiron** is verified by Herold, [p 327].

39. The 79th had been part of the "Dutch Expedition" that landed at Texel on 27 August, 1799, fought one battle at Egmont-op-Zee (2 October) and embarked for England two months later on 29 October, 1799.

40. General Jean-Baptiste Kléber (1753 - 1800), Bonaparte's successor, was assassinated by a fanatic in Cairo on 14 June, 1800, shortly after his stunning victory at Heliopolis (20 March, 1800). His successor as Commander of The Army of the Orient was General Jacques Menou (1750 - 1810).

41. Sir Ralph Abercrombie (1734 - 1801), born in Tullibody, Scotland, was related on his mother's side to the Dundas family. Due to his pro-rebel sympathies Abercrombie had not served in the American War of Independence. He tried law and politics and re-entered the Army at the age of 58, serving with distinction as a subordinate commander in several disastrous campaigns. In October, 1800, he had been given command of the Mediterranean, with particular emphasis on Dundas's proposed invasion of Egypt. Abercrombie was distrusted by many English officers because of his propensity to surround himself with Scottish officers.

42. "*The Camerons' Gathering*" was a name commonly used to identify the *piobaireachd* now known as "**Black Donald's March**" (***Piobaireachd Domhnuil Duibh***). *Piobaireachd* or 'pibroch' is the classical bagpipe music performed by highland pipers for hundreds of years.

43. Piers Mackesy [**British Victory in Egypt, 1801**] states that the Royal Navy lost 97 killed and wounded, the British Army 625. No figures are available for the French, but it is known they suffered less severely due to the protection from cannonading offered by the sand dunes.

44. "The Clan Cameron Feud" started in 1772 when Alan Cameron of Erracht (founder of the 79th in 1793) killed Alexander Cameron of Mursheirlich in a duel over the favors of a neighbor known as "The Red Widow". The dead man had apparently been egged on by John Cameron of Fassiefern, Factor to Cameron of Lochiel, the major branch of the Clan. Later, in 1791, Lochiel's Trustees instituted a court case against Erracht regarding a land purchase, a case which dragged on until 1811. Sir Ralph Abercrombie embroiled himself in the clan dispute in 1795 when his daughter married Cameron of Lochiel. This was two years before Abercrombie broke up the 79th while it was under his command in Martinique -- this despite the King's Letter of Service promising that the 79th would never be drafted to other units. The bitterness between the 79th and 92nd was even more enduring than the Clan Cameron Feud as can be seen in the literary output of members of both regiments over the next two centuries. As late as August, 1994, during the Edinburgh Military Tattoo this Editor discussed the merits of the two regiments with several serving soldiers of each, and discovered that an exceptionally bitter feeling still existed. Ironically, a month later the War Office amalgamated the two as one unit named simply "The Highlanders".

45. Captain Wyvill of the 79th mentions Sir Sidney Smith's commendation that afternoon, but makes no mention of MacHugh. [**Sketch of the Military Life of R.A.Wyvill**, London, 1820]

46. "That sailor is Sir Sidney Smith. He's a fancy dresser, but the sailors all say he's genuine. He shares in their manual work, and is as brave as a lion." 'Inky' was evidently a "practitioner of cant" (He spoke in London slang).

47. In 1799 while the 79th was stationed in Guernsey the regiment was made up of "273 Englishmen, 268 Scotsmen, 54 Irishmen, and 7 Foreigners" [**Historical Records of the Queen's Own Cameron Highlanders**]. MacHugh was probably included in the "7 Foreigners", due to his North American birth.

48. General Menou had previously become a Muslim in order to marry the daughter of a bath-keeper in Rosetta.

49. Smith's mission is also described by Sir Robert Wilson. [Mackesy, pp 149-150]

50. MacHugh was correct in his surmise. Hely-Hutchinson had been reluctant to agree to the flooding because of the damage it would inflict upon the impoverished population of the Delta. However, it would place an impassable barrier between his army and the mainland, and would enable him to leave a small holding force in front of Alexandria while he took the remaining troops eastward to Rosetta then up the Nile to Cairo. Today neither Lake Mareotis nor Lake Aboukir exist, a result of the constant growth of the Nile Delta.

51. The *skean dubh* (literally "black knife") was customarily worn in the top of the right stocking. It was a traditional highland weapon, rather than Army-issue.

52. MacHugh is correct in his general comments, but under-estimated the actual French garrison of Cairo that embarked at the end of the campaign. It totalled 13,672 soldiers plus 82 civilians. Not included were 3,000 or so women and children and 500 French who had deserted to join the Mamelukes. [Mackesy, p. 202.]

53. Downton was referring to a military flogging. It was standard procedure to employ the regimental drummers to flog those convicted of serious crimes although the 79th was noted for the rarity of such punishment.

54. D. Robertson describes this incident in his **Journal of Sgt. D. Robertson, late 92nd Foot: Comprising the Different Campaigns between the Years 1797 and 1818** (Perth, 1842) p. 18-19. Robertson mentions that the 92nd's party also brought along hammers to use in taking souvenirs, so it would appear that even then British soldiers were keen souvenir-hunters and tourists.

55. The officer was probably Captain Robert Dalrymple of the Third (Scots) Guards who mentions the incident in his diary, describing Colonel Cameron as "a blunt, honest Scotchman, but a great Goth". [Mackesy, p 197]

56. The gender of the Sphinx was a matter for debate until 1816 when excavations discovered the remnants of a beard which had been removed centuries previously for unknown reasons. It seems that men preferred to think of such a powerful image as a male, while the ladies were convinced such a mystical and enchanting face had to be female.

57. In 1804, Cameron sent to the Commander-in-Chief, The Duke of York, an "engraved Piece of Granite having formed part of the Sarcophagus in the Great Pyramid near Ghiza in which the <u>Founder</u> of that <u>Ancient Pile</u> is supposed to have been deposited." The Duke of York thanked him for "His Attention in sending Him so interesting a piece of Antiquity." [Loraine MacLean, **Indomitable Colonel**, pp178-9]

58. Poor Zenab! Few biographers of Napoleon acknowledge the sixteen-year-old despoiled by the married General. The girl's fate is mentioned only in passing by Christopher Hibbert who quotes Abd el-Rahman el-Djabarti: "Zenab had been debauched by the French. The Pasha's emissaries presented themselves after sundown. They brought her and her father to court. She was questioned about her conduct, and made reply that she repented of it. Her father's opinion was solicited. He answered that he disavowed his daughter's conduct. The unfortunate girl's head was accordingly cut off." [**Napoleon: His Wives and Women**, (London, 2002) p. 87]

59. Although details of the massacre itself are scarce, the aftermath including the subsequent rescue operation was reported by several eye-witnesses. The prisoners taken on the night of 20-21 October in Cairo were released under pressure from the British government. Only two of the Beys "arrested" at Aboukir survived.

THE MacHUGH MEMOIRS ～ (1798 - 1801)

MUSICAL COMPOSITIONS
BYPIPER R.G. MacHUGH

THE SWEET PINE HILLS   Slow Air

RG MacHugh, 79th CAMERONS, Giza, Egypt: June 29, 1801.

~ A SECRET OF THE SPHINX ~

OUR OLD CIAMAR A THA THU RG MacHugh, 79th Camerons

For 'Old Ciamar A Tha Thu' Lt. Col. Alan Cameron of Erracht, 20 October, 1801.